AF837399

Business and climate policy

Business and climate policy: The potentials and pitfalls of private voluntary programs

Edited by Karsten Ronit

United Nations University Press

TOKYO • NEW YORK • PARIS

United Nations University Press
United Nations University, 53-70, Jingumae 5-chome,
Shibuya-ku, Tokyo 150-8925, Japan
Tel: +81-3-5467-1212 Fax: +81-3-3406-7345
E-mail: sales@unu.edu general enquiries: press@unu.edu
http://www.unu.edu

United Nations University Office at the United Nations, New York
2 United Nations Plaza, Room DC2-2062, New York, NY 10017, USA
Tel: +1-212-963-6387 Fax: +1-212-371-9454
E-mail: unuony@unu.edu

United Nations University Press is the publishing division of the United Nations University.

Cover design by Andrew Corbett

Printed in the United States of America for the Americas and Asia
Printed in the United Kingdom for Europe, Africa and the Middle East

ISBN 978-92-808-1214-5

Library of Congress Cataloging-in-Publication Data

Business and climate policy : the potentials and pitfalls of private voluntary programs / edited by Karsten Ronit.
p. cm.
Includes bibliographical references and index.
ISBN 978-9280812145 (pbk.)
1. Business enterprises—Environmental aspects. 2. Environmental policy. 3. Climatic changes. 4. Voluntarism. I. Ronit, Karsten.
HD30.255.B8697 2012
363.738'745765—dc23 2012008153

Printed and bound in Great Britain by
Marston Book Services Limited, Didcot

Contents

Tables and figures

Tables

Figures

Preface

Climate change is an important issue in global politics but it is often overshadowed by more immediate concerns that reach the headlines and demand instant action. The issue of climate change has, however, undergone an interesting development from being a problem debated almost entirely by highly specialist scientists a few decades ago to one that is today addressed by a range of political decision-makers as well as civil society groups and the business community.

Many policy tools are available in the development of climate policy. This book analyzes the contributions made by business aimed at halting climate change and reversing its anticipated dangers. It is, however, difficult to predict how changes in the short and middle run will ultimately affect solutions in the long term, and several factors must be taken into account. However, a principal concern is the aligning of corporate goals with those of society. There is no uniform answer from the business community and no coherent strategy, but many firms and industries have turned challenges into opportunities and adopted a variety of private voluntary programs. These programs do not apply a single formula but are very much characterized by the different economic, organizational and political properties of the various industries and the readiness of firms to engage in new forms of cooperation. Our experiences with private voluntary programs, therefore, are promising but also, in some respects, rather daunting; all these experiences are valuable and can inform research on other sectors of business.

The project on business and climate policy, the results of which are presented in this book, began life in 2008 when I received a research grant from the University of Copenhagen and managed to put together an international research team. From the outset, everyone who has participated agreed happily to join and share their expertise on different sectors of business and their involvement in climate policy. To strengthen our cooperation, we have been in very frequent contact the whole way through, and to set the project on the right track from the beginning and to follow up at later stages I have undertaken two research stays: I was a visiting scholar at Yale University in 2008 and at the University of Washington-Seattle in 2010.

The first drafts of the chapters were presented at a workshop in Skodsborg, Denmark, in 2009, and on that occasion we had the opportunity to discuss initial thoughts about achievements in specific industries as well as the general framework to guide our research. Second and revised drafts of the papers were presented at a panel of the annual convention of the International Studies Association in New Orleans, in 2010. Here, the almost-final results of our findings, and how they could be compared and summarized into a general picture of private voluntary programs in climate policy, were on the agenda. In this context, comments and suggestions from Mark A. Boyer and Steven Bernstein, our discussants at the panel, were very helpful.

In the final stages, the book manuscript has further benefited from valuable comments from three anonymous reviewers. Final thanks go to Naomi Cowan, editor at the United Nations University Press, for swiftly guiding the manuscript through the different stages of the production process.

Karsten Ronit

Contributors

Frederike Boll is a PhD candidate at the Westphalian Wilhelms University, Munster, Germany.

Laura Bozzi is a PhD candidate at the Yale School of Forestry and Environmental Studies, Yale University, USA.

Benjamin Cashore is Professor in Environmental Governance and Political Science, Yale University, USA.

Jennifer Clapp is a CIGI Chair in International Governance and Professor in the Faculty of Environmental Studies and Balsillie School of International Affairs at the University of Waterloo, Canada.

Doris Fuchs is Professor at the Department of Political Science, at the Westphalian Wilhelms University, Munster, Germany.

Lily Hsueh is a PhD candidate in Public Policy and Management at the University of Washington, Seattle, USA.

Kelly Levin is senior associate at the World Resources Institute, Washington DC, USA.

Constance McDermott is a James Martin Senior Research Fellow in Forest Governance, University of Oxford, UK.

Tony Porter is Professor of Political Science at McMaster University in Hamilton, Canada.

Aseem Prakash is Professor of Political Science and Walker Family Professor for the College of Arts and Sciences, at the University of Washington, Seattle, USA.

Karsten Ronit is Associate Professor at the Department of Political Science, University of Copenhagen, Denmark.

Jason Thistlethwaite is a PhD candidate in the Global Governance Program at the University of Waterloo, Canada.

Part I

Introduction

1

Marrying climate policy and private voluntary programs

Karsten Ronit

Introduction

Climate policy has experienced a fast and fascinating development over recent decades, but the policy is still vastly underdeveloped and immature when measured against the overwhelming tasks with which it is expected to cope. Concerns about climate change are today making headlines across the world, and conflicts arise on how to interpret related data correctly, to design timely and effective strategies for coping with climate change, and to make institutions work. Thus, all kinds of decision-makers and policy advisers are – notwithstanding their strong commitments and efforts – relatively inexperienced in the field of global climate policy and new and complex forms of policy coordination are needed.

In the public realm, states and intergovernmental organizations in the UN system and beyond have invested significant power in formulating and negotiating appropriate strategies to halt climate change, which entails many dire consequences for both North and South. It is a task that must involve various established organizations that were not originally designed to address these issues per se but that are now integrating various climate components into their agendas. Climate change is not merely an extension of environmental concerns, although agencies specialized in this field are particularly active. Indeed, many intergovernmental organizations observe various aspects of climate change from a rich diversity of angles,[1] indicating a certain fragmentation of efforts (Aldy and Stavins 2010; Axelrod, Downie and Vig 2005; Biermann and

Business and climate policy: The potentials and pitfalls of private voluntary programs, Ronit (ed.), United Nations University Press, 2012, ISBN 978-92-808-1214-5

Bauer 2005; Biermann, Zelli, Pattberg and van Asselt 2010; Chambers and Green 2005).

Dynamic activity has also required much innovation in the form of new forums and new initiatives. Concerns about climate policy go back a long way; however, a major breakthrough was achieved at the Earth Summit in Rio, in 1992, from which the UN Framework Convention on Climate Change (UNFCCC) emerged. Today this convention is administered by a UN secretariat carrying the same name. This process also led to the Kyoto Protocol, which is now being followed up and renegotiated around the Conference of the Parties (COP), which organizes a special series of meetings. Parallel efforts have evolved around the Intergovernmental Panel on Climate Change (IPCC), established in 1988 by the World Meteorological Organization (WMO) and the United Nations Environment Program (UNEP) to scientifically assess human-induced climate change, with a focus on the impact of greenhouse gases on our climate. The IPCC delivers key inputs into the political process, and its scientific reports inform climate negotiations in many ways. Major concerns in these debates are the timely recognition of threats to humankind, the targets to be set for emissions, and ways to create appropriate institutions to meet these challenges.

However, the potential of regulation is not exhausted through the accomplishments of states and intergovernmental organizations. Neither the various obstacles that have been overcome and the successes achieved, nor the failures and unfulfilled promises of international negotiations, rule out the active role of other interested parties. These efforts, however, are often neglected in the current debate. Alongside conferences of states and the continuing work of international secretariats, various groups of private actors have emerged. Indeed, it is one of the characteristic features of this emerging policy field that critical civic groups, scientific experts and pioneering firms have been decisive in bringing climate change onto the political agenda (Carpenter 2001; Newell 2000).

Nonetheless, their role is not simply to provide inputs to traditional decision-makers. Private actors are also capable of developing and administering private schemes intended to halt climate change; chapters in this book capture some key features of these diverse initiatives. Essentially, *private voluntary programs* are *private* in the sense that they are initiated and joined by business, but some programs are sponsored by government and various stakeholders outside the business community and reach out to firms and industries. They are *voluntary* in the sense that business is free to join or leave them, but various kinds of moral, economic and political pressures exist in business and in the surrounding environment and contest their voluntary character. Finally, they are *programs* in the sense that a variety of formal rules and bodies may be established to administer and evaluate the schemes, but the stringency of programs varies: some programs are challenging for business to comply with, while others are weak.

Sometimes "voluntary programs" are viewed exclusively as government-sponsored programs which firms are free to join (Baranzini and Thalmann 2004; Morgenstern and Pizer 2007). In this book private voluntary programs embrace a broader category of arrangements. This perspective builds on a tradition for studying the policy instruments of private actors in comparative and international contexts in which private actors provide important alternatives to traditional top-down public regulation, but usually face a number of enforcement problems.

Accordingly, private initiatives can be sponsored by many actors, but business is undoubtedly the most vigorous force (Cutler, Haufler and Porter 1999; Ronit and Schneider 2000). These efforts do not exclude actions by civil society, expert communities or public institutions. Important encouragement comes from outside business groups, and occasionally these forces even develop schemes and offer them to business.

This argument does not suggest that private schemes are necessarily superior to measures agreed between states and that they can totally replace them; rather it indicates that there is a problem-solving potential, the prospects and limits of which deserve independent scrutiny. Thus, private efforts become part of an emerging "governance architecture" in climate policy at a time when public strategies are in the making and public institutions are in the process of being formed, creating special opportunities for private regulation. This "architecture", it is argued, includes not only the traditional perspectives on states and intergovernmental organizations but also a variety of other actors and their relationships (Biermann et al. 2010). This perspective, of course, requires that we focus attention on each of the actors, including firms and industries as key private actors. Consequently, the overriding questions are these: In what respect can private programs be married with public concerns in relation to climate change? Do private actors have special competencies that make them particularly suitable for this task? And under what conditions are such programs likely to evolve, succeed or fail?

In the following section of this chapter we identify and discuss the major characteristics of climate policy, which has materialized as a unique policy field, and we seek to understand whether any of its features are likely to complicate or support private contributions to problem-solving. We further interrogate the debate on private regulation to examine how initiatives have evolved and to highlight some of the advantages and shortcomings typically associated with private regulation. Although private regulation deserves to be studied in its own right, such study does not suggest that private voluntary programs are decoupled from other actors, and we discuss how these other actors and the conditions that they shape may impact upon business behavior. This is an important aspect in the development of the study of private voluntary programs, because various barriers and facilitators can be found within the business community itself, and also in the ways that business interacts with public institutions, civil society organizations and expert groups. To assess the role of these programs, it is moreover important to uncover the concrete mechanisms in their emergence and administration, so we identify core elements such as designing, joining, monitoring and complying as key parameters to be analyzed.

Finally, this chapter closes with a brief description of the studies in this book. They focus on experiences from programs that embrace multiple sectors or particular industries. By drawing on experiences across the business community and offering in-depth studies of sectors, each having a particular record of performance in relation to the emission of greenhouse gases, we identify some of the major factors that facilitate or impede the introduction of private programs in business and, thus, offer perspectives to understanding the potentials and pitfalls of private programs in climate policy.

The public dimension and the goals of climate policy

Climate policy is a complex field that is linked with environmental policy in a broad sense, has a close relationship with energy policy, and has some connections with other areas such as ozone policy. Taking an even wider perspective, the policy is related to a range of demographic and economic issues. But as an object of enquiry it has gained its own profile and embodies some distinctive issues. Thus, the focus of current climate policy is decidedly on greenhouse gases as the chief source of climate change, with carbon dioxide emissions in the foreground. A key question permeating many analyses is, consequently, how governments and international agencies gather and interpret scientific data, address emission issues, negotiate and design effective strategies, and employ specific regulatory tools.

At the same time the use of private regulation across an increasing number of problems has led to a new and specialized strand of research presented under different concepts, but all having an interest in regulation beyond the typical top-down approach under the auspices of public agencies (Doh and Teegen 2003; Hemmati, Dodds, Enyati and McHarry 2002; Pattberg 2007; Ronit 2007). The role of private voluntary programs, as examined in this book, is located within this broader tradition of private regulation.[2]

Various forms of private regulation are pertinent to the solving of economic, social and environmental problems, and a central task is to ascertain which programs actually host sufficient capacity to offer a relevant alternative to public regulation, and which programs underperform. It should be noted, however, that although private regulation has been put forward as a particularly relevant device in public policy over the last decade and during a period when climate change issues have also come to the fore in global politics, climate policy and private regulation have not been systematically brought together and the possible benefits have not been sufficiently appreciated. However, some clues as to how this particular form of regulation is able to handle climate problems are offered in existing research, and we discuss policy-oriented studies on climate issues that take up the opportunities put forward in private programs and also examine whether climate change appears particularly fruitful – or unproductive – to address from the perspective of private regulation.

In the growing body of literature on global climate politics, two broad and related approaches stand out: state-centered and policy-centered analysis. First, a variety of studies grapple with the role of states in relation to the environment in general and climate issues in particular (Luterbacher and Sprinz 2001). States are often regarded as self-interested actors that formulate strategies and negotiate treaties around climate policy, with the outcome very much determined by patterns of conflict and cooperation; but state-centered approaches also accommodate constructivist thinking in which the priorities given to climate challenges are more arbitrary and open to interpretation (Harrison and Sundstrom 2010; Pettenger 2007).

Along with the development of climate policy as an emerging policy field and the creation of new and specialized global forums, attention has been directed towards intergovernmental organizations, regimes, and the various activities surrounding them. Indeed, these forums not only provide an arena for disputation between states but also gain some independence. Some special agencies have a wide reach and inform and influence the processes of climate politics, calling into question the dominance of state-centered perspectives, although the focus remains on public forms of authority.

In some respects these perspectives are relaxed when combined with a stakeholder perspective. Thus, attention is extended to private actors making inputs into agenda-setting and in the formulation and implementation of public policy (Orr 2006; Raustiala 2001). Climate politics is consequently recognized as a multi-actor and multi-scale problem (Bulkeley and Newell 2010: 2–5). Different analytical emphasis can be put on the roles performed by either public or private actors at different levels; in principle, there is even the possibility that leverage can be so strong that public institutions are captured by private interests (Fuchs and Vogelmann 2008).

Second, a related strand of research is occupied with the very content and impact of public policy. Climate stability is a basic global public good, and halting climate change demands public solutions on a grand scale; hence, solutions are linked to the active involvement of public institutions. Accordingly, states and intergovernmental organizations are instantaneously singled out as the primary players in the field of climate policy because they are supposed to host the primary, best or perhaps even sole potential for providing public goods. Public goods are characterized by their non-rival and non-excludability status, but the provision of public goods is still a complex issue (Cornes and Sandler 1986) because "climate" also entails several public-good dimensions in relation to territory and time (Kaul and Le Goulven 2003).

As to the dimension of territory, policies must be truly global because of the transboundary effects of greenhouse gases. Climate belongs to the broader category of "atmospheric commons" (DeSombre 2006: 97–128), and all people enjoy the same atmosphere, making climate a really large-scale public concern. In this context, it is, however, important to understand the specific character of public goods and the linkages between commons at different territorial levels (Keohane and Ostrom 1995). To varying degrees, all countries and regions are affected by climate change; however, those areas most seriously impinged by current or future changes are not necessarily those whose production models, consumption patterns and lifestyles more generally cause climate change, all factors that demonstrate that climate policy is essentially a global issue. To design appropriate policies we must realize how these levels are linked because changes for the better – or the worse – in one part of the world not only have local impacts but may even affect the whole system. Thus, endeavors to reduce CO_2 emissions in the industrialized countries and in the major economies are crucial; however, efforts in developing countries are also needed, such as avoiding the destruction of fragile biospheres by deforestation.

As to the dimension of time, policies must be designed in such a way that they lead to the adaptation and mitigation of climate change so that future generations will be able to enjoy the benefits of favorable climate. The reasons for the current problems in global climate are attributable to changes accumulated primarily over the last century, and appropriate policies adopted now will have an impact on future generations, although the true effects of the policies will take several decades to materialize. Most alarming perhaps are the situations where changes in climate will simply destroy habitats and make life impossible, but challenges are being presented by many other issues that in the short or long run require solutions. Indeed, climate change problems are often characterized as being in need of urgent action to avoid future disasters. This will, for instance, include important private and public learning processes, a theme picked up by Laura Bozzi, Benjamin Cashore, Kelly Levin and Constance McDermott (Chapter 4).

Consequently, this intergenerational side of policy must be given serious attention and is addressed in the public-goods literature (Kaul, Grunberg and Stern 1999). It seems that the market alone cannot handle these challenges and that the "polluter pays" principle, first adopted at the 1972 UN Conference on the Human Environment, is not applicable to today's climate policy. Basically, the principle suggests that social costs of production can be avoided when firms internalize costs and the price mechanism is used to reflect the real social costs of production, for which the consumer pays. This path, of course, has an impact on the competitiveness of firms in relation to climate policy, creates different incentives to climate-friendly production and offers different opportunities for firms to establish and sign up to programs. But it is important to note that it takes decades to accumulate meaningful effects on the climate and, hence, it is difficult to hold particular firms accountable for past misconduct. Furthermore, some damages to the climate are simply irreparable: they cannot be reversed once critical "tipping points" are passed.

Given the potent role of states and intergovernmental agencies and the many and complex dimensions of climate policy, it seems obvious that climate change must be addressed in a public framework; however, climate negotiations have faced many difficulties: interests often differ, free-riding can occur, and mutual trust may be low. Efforts to provide public goods, however, are also discussed in literatures that not only are concerned with the actual or potential role of public institutions but see a number of other actors, including business and civil society, as highly relevant contributors to global public policy (Reinicke 1998), an important issue in recent years. Consequently, we must consult another body of research to examine whether problems, such as climate change, can be mitigated beyond public institutions and under what circumstances corporate actors alone, or perhaps in concert with civil society and expert communities, have the required potential to facilitate problem-solving.

Climate policy and business: A perfect misfit?

Both scholarly and practical interests in private arrangements have grown considerably, and experiences have been gained from a rich diversity of policy fields (Graz and Nölke 2008). It is also clear that business has a variety of economic and political roles to play in relation to climate change (Newell and Paterson 2010; Pinkse and Kolk 2009), only some of which, however, take the form of private voluntary programs discussed in this book. To get a better grasp of these alternative approaches, we here distinguish three analytical and interrelated perspectives in existing research which can also be related to climate change: firm initiatives, sectoral, multisectoral and cross-sectoral schemes, and outside-based arrangements. These levels are discussed in relation to a number of cases but have thus far not been systematically applied to climate issues – although some include climate change.

First, a significant part of the literature is focused on single firms adopting their own independent strategies to meet regulatory challenges in various domains (Vogel 2005), including climate policy. Corporate actions are based on either commercial concerns or societal motivations, or a blend of both. It can be difficult, however, to properly discern whether these actions are driven by self-interested rationales or triggered by ambitions to solve societal problems. The environmental field, for instance, provides some evidence. Parts of the literature here see these as basically contradictory strategies and are more confident that public regulation and an active civil society will produce better solutions (Lipschutz 2005; Smith 2008), whereas other contributions analyze in which respects different motivations can best be aligned and how corporations take upon themselves new responsibilities and adopt new roles (Crane, Matten and Moon 2008; Prakash and Potoski 2006; Schmidheiny 1992).

The only feasible solution for firms may be to aim to gain the competitive edge and improve their reputation in markets that demand and appreciate a wider corporate social profile. Such strategies can be launched relatively autonomously by firms that adopt their own codes of conduct and put their own principles into practice (Vogel 2005). This strategy requires the installation of effective in-house mechanisms to monitor and report on their own behavior, and correct it if necessary. The role of corporate pioneers can be crucial because firms can avail of some benefits in the early stages before other firms follow suit and a new level of performance is achieved (Gunningham and Sinclair 2002; Porter and van der Linde 1995). At a time when climate change is coming onto the agenda, the situation is ripe for business to adopt climate-friendly strategies, and economic interests may here chime with public goals. However, for such single-firm norms and rules to emerge, the organization is often under the strong and perceptible influence of competing firms at the same level or at different levels of the production chain; in addition, customers and other interested parties or firms may find it relevant to band together with other firms to raise standards. We here move into other traditions of studying collective rule-making in business.

Second, the part played by business collective entities in rule-making is scrutinized in another strand of research on self-regulation (Streeck and Schmitter 1985). The perspective on collective action is applied across a large group of industries and product categories, including very large entities, in our case, for instance, the World Business Council for Sustainable Development (WBCSD) or the International Chamber of Commerce (ICC), which, however, also encourage single-firm action. The climate challenge has also created a demand for new forms of cooperation in addition to domain expansion of existing business (Bulkeley and Newell 2010: 87–104). When we move from rules applying to individual firms to rules applying to groups of firms, these rules must be negotiated and agreed on between independent firms. In other words, the adoption of many standards becomes an extension of competition in the market, because the level of regulation is likely to benefit some firms and bring competitive advantages, a point we shall later elaborate. In other words, we need to understand single firms in a group context, and so different literatures must be connected.

These broader arrangements are driven by industry associations or special clubs and other collective outfits, and they employ different mechanisms to consolidate arrangements.[3] Indeed, some business associations are based on a commitment to organize firms within a given interest category in an encompassing and inclusive fashion. Although these goals are not always achieved, standards defined by business associations are typically more ambitious in terms of collective action and superior to arrangements covering a smaller subset of firms.

Some entities operate with a form of invited membership; they apply selective membership criteria to exclude underperforming firms and to guarantee that the entity is associated with higher standards. In the same way, clubs to which firms must themselves apply for membership and must meet certain criteria to become members – if they are not already among the recognized founders – exclude certain firms in order to keep standards high. Provided these different kinds of collective entities, backed to varying degrees by the active promotion of individual member firms, are able to clearly distinguish members from non-members, members enjoy various reputational benefits. In this context, the dissemination of relevant information about the special contributions to societal problem-solving, in our case climate change issues, can become an important vehicle to promote arrangements.

Third, we must draw attention to approaches that conceptualize and analyze standards established either on the edge of the business community or beyond it, embracing multiple stakeholders (Bendell 2000; Pattberg 2007: 91–96; Yaiji and Doh 2009). In some sectors this novel form of rule-making has become important and has been well scrutinized in a small but expanding literature. This is a qualitatively different form of rule-making than single-firm or industry-based rule-making because authority is now shared with, or in some cases even surrendered to, other stakeholders. It is still private regulation and is thus very different from regulation adopted by public institutions, but it is not limited to business influence, and even government agencies can be involved, as the Climate Group, an alliance of companies and government institutions, demonstrates.

Again, and with the risk of oversimplifying complex patterns of regulation, we may single out two major directions in these literatures that in fact reflect diverse practical experiences with private regulation, namely, standards defined between business players and external parties, and standards sponsored exclusively by these external parties in the form of civil society actors or expert bodies. In both cases, however, these private initiatives have as a very basic condition that corporations and industries join them and commit themselves to their goals.

As to standards established through cooperation between business and external actors, a further distinction can be made between cooperation between either single firms and various external groups, or collective entities and these same groups. Again, those more encompassing rule systems that are brought to life through cooperation between industries and various affected stakeholders, and that are supported by these, can be characterized as having a greater problem-solving capacity than those involving single firms. However, if the latter firms are very large, cover a significant proportion of an industry, pioneer solutions, or, in the case of climate change, represent an important initiative, then dialog with single firms and recognition of their initiatives can be of great value. Tony Porter investigates these issues in his chapter on the automobiles sector, which is characterized by a high degree of concentration and a high visibility of players in the form of car manufacturers.

As to standards established through arrangements under some kind of influence from the business community or perhaps designed primarily by forces outside it, civil society organizations and expert groups can be particularly innovative, launching relevant schemes before appropriate action is taken in business. These actors can – with or without the formal cooperation and consent of business – formulate important rules and create relevant bodies that are open for firms to participate in, provided they meet certain membership criteria and comply with the rules of a given initiative. Becoming a member implies that firms or specific products are certified, and firms can often better present themselves as responsible corporations when recognition comes through a not-for-profit entity.

Private rules are offered at various levels in the business community, but a tendency prevails in much current research to see climate change problems as most appropriately solved by public sector institutions and through traditional top-down public regulation. Public sector institutions are in many ways more visible than private sector bodies, they are fewer, and in many ways they lend themselves more easily to coordination than the "private world". In this context, space is left for the analysis of business and other private actors in influencing climate policy (Levy 2005), and these groups are recognized as leveraging states and international organizations.

That private actors, and in particular business, do launch effective programs is often ignored, often from the belief that business is creating problems rather than solving them. Several critical voices – and different shades of criticism exist – have pointed out some of the principal problems in business taking on public tasks (Conzelmann and Wolf 2008; Newell 2005), and some go into a further discussion on the implementation of voluntary private programs with regard to the environment, and climate more specifically (Baranzini and Thalmann 2004; Clapp 1998; Morgenstern and Pizer 2007; Vogel 2005: 121–132). Further, they discuss the difficulties of establishing sound criteria for successful performance and extending these programs to achieve a global status. Most typically, these studies argue that there are various deficiencies in private programs and that they cannot stand alone; however, we need a clearer understanding of these shortcomings.

Although some studies on private regulation endorse the view that business is incapable of handling public policy issues, and typically castigate efforts at the level of single firms and industry bodies, we also find research that is more confident in bringing business into problem-solving as a relevant alternative to public regulation. Accordingly, climate policy and business are not necessarily a misfit. Under specific circumstances corporations, industry associations and other collective entities are, it is argued, in a position to use their self-interested behavior to achieve public goals. Different emphasis is, however, attributed to the role of firms and collective entities, and different views prevail as to the necessity of dialog and formal cooperation with other interested parties; but after all, these disagreements do not block but rather espouse a joint effort to empirically investigate the prospects of private voluntary programs in climate policy. If, indeed, climate issues are particularly difficult for private regulation to handle because of their strong public-good character, positive signs and possible achievements of private regulation in this field will be especially interesting.

Institutional conditions: Barriers and facilitators

Major responsibilities in private regulation lie with business. Ingenuity and resolve are essential to the creation and implementation of schemes that become valid in business and are recognized in society. However, the development of schemes also hinges on a variety of institutional conditions, because factors beyond the control of the business community bring various pressures and encouragements to bear upon firms, and these can impede or facilitate private programs. In various ways, these factors are at play in relation to issues such as: How do programs relate to other rule systems? Who can become members and what standards must they follow? Who controls their correct implementation, and who decides on action in cases of non-compliance? And what ends should programs achieve? We identify four clusters of relationship and institutional conditions that business must cope with, and which challenge business in different ways.

First, corporations interact with each other in the development of programs, with corporations being both competitors seeking comparative advantages and potential partners seeking common ground when so demanded. Second, corporations relate to public institutions, because business takes action to avoid imminent threats of public regulation and finds a role in cases when public institutions encourage business to fill a void by designing private forms of regulation. Third, corporations engage with civil society organizations that represent consumers in the market or critical voices in the public discourse, and find relevant forms of dialog and cooperation. Fourth, corporations listen to and use knowledge disseminated by various expert organizations to consolidate private programs, and formal cooperation may arise to sponsor and administer joint programs. These different conditions, and the challenges and opportunities they generate, will now be further discussed; we will also show how we must account for them when analyzing private voluntary programs and their application to climate policy.

As the first conditioning factor, we denote internal processes in business, which are of primary importance both when corporations single-handedly draw up rules for themselves, and when norms and rules are coordinated between a smaller or larger group of firms (Delmas and Montiel 2008: 88; Parker 2002; Prakash 2000). In Chapter 5, Doris Fuchs and Frederike Boll pay particular attention to the incentive structures facing firms and the various mechanisms driving their calculations. Firms typically watch the behavior of competing firms in the market, and in society more generally, and they are keenly aware that the strategies of other firms can have a huge impact on their own competitiveness. First movers adopting new practices in relation to climate change will, under the right circumstances, enjoy some benefits; however, they run certain risks if such practices are not observed and appreciated. Embarking on new practices in relation to climate change may be easy if a firm has already taken some recognized initiatives in other economic, social or environmental domains and has some experience in handling codes of conduct and other rule systems.

It can be a key objective of firms to inspire and win other firms for the cause – in fact, the poor reputation of an industry is likely to cast a shadow even over innovative firms. If successful, pioneers can be followed by other firms that design similar rules, but in such cases there is also a certain probability that joint rules regulating the behavior of a larger group of firms will be formulated and implemented. Indeed, this may take place in the context of a collective entity. In this way, we can analyze firms as being able not only to capture public policy (Stigler 1971) but also to capture private organizations regulating various aspects of corporate behavior, because the levels and forms of regulation are disputed and because who controls private organizations is an important parameter of competition.

In many economic, social and political cases, firms coordinate (Cutler, Haufler and Porter 1999; Gunningham and Rees 1997; Hollingsworth, Schmitter and Streeck 1994; Ronit and Schneider 1999). However, not all firms within a given interest category are necessarily involved in or are asked to participate in such an enterprise. Business may see many possibilities in adopting voluntary programs, but there is no uniform attitude among those corporations that are confronted with intricate climate issues – the chapters of this book provide evidence of this. Indeed, we may anticipate that some firms will not be able to meet a common set of criteria, nor have any intention of doing so. Although free-riders can be highly problematic and may damage the reputation of compliers, and it can therefore be important to exert pressure on and recruit them, there are many costs associated with their inclusion. Consequently, a number of complex dilemmas emerge.

It is important to note that in cases of collective action, the participating firms are not necessarily the only ones responsible for making private regulation work. Very often a business interest association is assigned such tasks. Collective action may be facilitated if an association or some other kind of collective outfit already exists, and if the firms concerned do not have to organize around a particular issue, such as climate change, from scratch. In other words, business associability can be a key factor in fostering cooperation; we will examine how this process has conditioned the development of private voluntary programs.

As a second conditioning factor, public institutions and strategies can have a huge impact on private regulation (Vogel 2005: 132), showing that corporate action is not stimulated only by various intra-business variables. Much of the literature on climate policy, however, almost takes public regulation of climate change issues for granted, and business is up against a strong regulatory trajectory when trying to establish private regulation. From the perspective of some public decision-makers, it may seem a contradiction in terms for business to take charge of a global public issue such as climate change, but we envisage that government agencies and intergovernmental organizations can provide a basis for corporate action through voluntary agreements entered into by business and government, and stimulate partnerships (Pinkse and Kolk 2009: 41–62). Authority is fragmented (Biermann, Pattberg, van Asselt and Zelli 2009) but various soft law measures, including several initiatives in the private sector, can indeed coexist with hard and traditional law adopted by public authorities (Kirton and Trebilcock 2004).

General initiatives such as the UN Global Compact, praised by some and criticized by others, can also have some relevance with regard to climate change issues, and it has been most manifest in the Compact's offshoot, Caring for Climate, created in 2009. Therefore, private regulation may be stimulated by public institutions in a number of situations – either because public institutions surrender authority or because they sponsor programs. It is difficult to conclude, however, that these situations are invariably related to "weak governments" (Prakash and Potoski 2006: 19). If negotiations between states are crowned with success, business may be facing an opportunity to take over some aspects of enforcement because there may still be tasks for business to perform once the big political conflicts are overcome.

Indeed, a useful distinction with regard to different policy instruments with a potential to involve business and other private parties has been made between compensatory approaches and innovative strategies (Pinkse and Kolk 2009). If public institutions fail and if negotiations are slow and lead nowhere, a number of opportunities for business may arise, and resources missing in the public domain can be invoked in business. There are also political and technical problems, though, with implementing different mechanisms in business, for instance, emissions trading (Victor 2001: 55–74). In fact, the demand for private regulation and its specific organization may be motivated by achievements as well as failures in the public system.

There is, however, a greater likelihood that governments and international organizations are inclined to support the emergence of voluntary programs when these become a supplement rather than a full alternative to traditional public regulation. Studies with a regime focus on environmental accords also stress, for instance, that different levels of regime effectiveness apply (Miles, Underdal, Andresen, Wettestad and Skjaerseth 2002), and it is likely that alternative policy instruments are considered when public institutions are relatively weak. Indeed, various forms of self-regulation in the shadow of the state can be envisaged (Boddewyn 1985; Gunningham and Rees 1997).

New bodies can, of course, be surrendered authority from public institutions by design when it is realized that public action alone will fail, or at least will not be able to effectively take on those tasks that are assigned to new bodies. Indeed, this provides a key background for the emergence of a range of programs in the US, as will be discussed by Lily Hsueh and Aseem Prakash (Chapter 3). There can be various formalities associated with such a process, and the wheeling in of business is based on different kinds of mutual understanding. Thus, public institutions can take various courses of action through explicit encouragement; what is important is that we see the active involvement of business as potentially influenced by top-down processes in which elements of actual or potential authority are surrendered.

Nevertheless, business is not sitting on its hands waiting for a message from governments or international agencies, and different paths leading to private regulation in the context of public influences can be envisioned. Indeed, business may take authority with or without the prior consent of public institutions to avoid the various risks associated with public intervention (Haufler 2001; Vogel 2005). Although many strong pressures emanate from governments and international agencies, it would be wrong to underestimate very important bottom-up processes in business. As discussed earlier, there are many reasons for corporations, associations and other collective entities to engage in rule-making, and in initiatives in which competitiveness is decisive. However, we should not forget the power component and the fact that business also influences the demand and supply of private regulation (Fuchs and Vogelmann 2008). It would also be easy to underestimate the significant innovative potential offered by firms alert to different signals in the marketplace, if we ignore the fact that business can observe demands in the market before public institutions and can integrate public and private goals in climate-related programs. These programs, then, can at a later stage be recognized by public institutions if these are in line with public ambitions in climate policy.

A third conditioning factor in the environment of business lies in the role of civil society. As with public authority, this role can be performed in many different ways. The global Climate Action Network (CAN) is a particularly interesting organization addressing climate issues,[4] but other specialized entities are also orientated towards particular sectors of business. Civil society organizations often have a moral authority, and influence the public discourse in ways that business must find appropriate answers to, and, if necessary, change its patterns of behavior – corporation by corporation, or through coordinated responses. There is no doubt that the public discourse with regard to climate change has changed dramatically over the last decade and that various civil society groups have contributed to this. They leverage states and international forums, set new agendas, and in subtle ways influence business – for instance, through adverse publicity when the World Wide Fund for Nature (WWF) challenged firms and published a list of "Thirty Dirty" (30 least carbon-efficient power stations).

A lack of trust, however, prevails in large parts of civil society with regard to private regulation run by business alone. Serious doubt remains as to whether business can meet these challenges effectively and legitimately and as to which kinds of bodies are appropriate to solve these problems in business (Barber 1998; Bernstein and Cashore 2007; Coleman and Porter 2000). Instead, there seems to be a general preference in civil society for public regulation to halt climate change, and interestingly enough, many civil society organizations obtain some assistance from governments (Gunningham 2007: 211).

Although there is a lively public debate today on many aspects of climate change and a growing awareness of this problem, the interest does not necessarily trickle down to particular sectors. Special and targeted actions are needed for initiatives to evolve. It can be difficult, however, to trace definite influences of civil society and concrete responses from business unless they enter into a formal dialog and become involved in running joint programs. Civil society organizations may thus move from a policy of broad engagement to a strategy of determined participation and begin leveraging particular firms or industries in a more direct fashion, attempting a dialog or seeking formal cooperation if a clear climate component can be identified; in addition, new actors such as the US Coalition for Environmentally Responsible Economies (CERES), including business and civil society players, may become relevant.

Various dilemmas must be overcome because different views prevail in civil society as to the fruitfulness of such a strategy. These dilemmas are pertinent not just to each individual civil society organization but to the community of civil society groups as a whole, and different strategies are no doubt related to their self-confidence and capacities. In the complex field of climate change, we expect that it may be difficult to develop a strong capacity to match the business community, and, in this way, professionalize civil society organizations. In other words, business has to grapple with a set of changing perceptions and varying capacities in civil society. It is likely that business will interpret these challenges in different ways, as well.

Of course, we must also expect the reactions of business to the demands of civil society to be intimately connected with more mundane factors, such as realistic plans to reduce CO_2 emissions by changing existing forms of production, developing alternative products, using energy more sparingly and finding management capacities to monitor the behavior of firms and industries. If, however, a mutual understanding of the usefulness of cooperation comes about and the practical details of a program are worked out, civil society can become formally involved in the administration of regulation (Rondinelli and London 2003). An interesting type of challenge must be met, however, if civil society organizations and other non-business actors are to lead in the setting up of private schemes. In such cases, firms can typically participate on an individual basis or industries can participate through their associations, but these strategic situations are very different from various forms of business-driven programs; we shall elaborate this point later.

The institutional environment is made up of a fourth and final conditioning factor, namely the role taken by various sorts of scientific expert communities (Dessler and Parson 2006). In addition to regulatory, social and economic licenses to operate (Gunningham, Kagan and Thornton 2003), we must examine a scientific license to operate in the case of climate problems. It is necessary to distinguish these as a particular group separable from both public institutions and the civil society community by constituencies and functions, but doing this is not always an easy task. In fact, a major player involving key experts and expert communities is the IPCC, in which experts are nominated by governments but do not represent governments. Some groups, such as the Carbon Disclosure Project established in the UK, do not simply fuel the public debate but build upon a number of voluntarily "responding companies" and function as a third party disseminating information about their activities. The Global Trade Watch has a particular role to play in relation to emissions trading, with a special view to the conditions of developing countries.

A further group of organizations, about which there is limited research concerning their role in climate politics but which deliver significant inputs to the current debate, includes science-based entities and the translation of science into policymaking (Schroeder, King and Tay 2008). However, it is sometimes difficult to draw a line between science-based entities and those with close ties to business or civil society groups. An actor such as the business-related Pew Center on Global Climate Change.[5] for instance, has a hybrid character and sees itself as a non-partisan alliance between different stakeholders. And finally, various ad-hoc initiatives lead to scientific evaluations and reports, such as the acclaimed Stern Review.

Although autonomous, these experts are often linked to the work of a variety of public institutions, and there seems to be a preference for public regulation or for embedding private programs in public regulation across large parts of the scientific community on issues of climate change. The latest IPCC assessment report states, "Some corporations, local and regional authorities, NGOs and civil groups are adopting a wide variety of voluntary actions. The voluntary actions may limit GHG emissions, stimulate innovative policies and encourage the deployment of new technologies" (IPCC 2007: 29). The IPCC recognizes that voluntary arrangements have a role to play, though generally a limited one, and there is some ambivalence regarding this specific tool.

As we have discussed, scientists are granted a key role in international climate forums, the IPCC being a primary example, but experts are also indispensable and feed into many other intergovernmental organizations, including the International Organization for Standardization, and inform the policymaking of states. It would be strange, therefore, if scientific experts did not wield considerable influence on business, either through their work in these bodies or through their own independent scientific work. In fact, the public and business discourse around climate change is to a substantial degree informed by scientific analyses, and although certain specialized discussions are kept within the scientific community, many of them today cascade into the general debate and are picked up by business circles.[6] From a business point of view, there are consequently many and changing puzzles to explore, a fact that has implications for designing programs.

Expert knowledge, however, is not held entirely by external players; corporations and various collective entities typically have their own intelligence systems and build up their expert knowledge base. This expertise is employed in corporate contexts to achieve corporate ends, and it is used to professionalize private regulation. When developing and administering private schemes, for example, various kinds of legal and accounting expertise are needed to measure and secure the effectiveness of programs. Certain forms of expertise are also acquired from external and independent sources in auditing processes. They assist in the development and evaluation of rule systems and, through auditing, help monitor whether firms actually comply with standards and disseminate relevant information to business and to the general public. As will be detailed by Jennifer Clapp and Jason Thistlethwaite in Chapter 2 on the financial sector, this information is of overriding importance for investors. In the context of climate change, however, we anticipate that it can be enormously difficult to obtain relevant and timely data because problems are scientifically complex and because business players tend to develop a narrow expertise in relation to a particular industry and do not embrace broader climate issues.

Being important players, independent experts have many roles; they are active in public discourse, in providing inputs to political decision-makers, in delivering critical accounts of business performance, and in helping business build its own regulatory bodies. This activity has, in part, been dealt with in the epistemic community literature (Haas 1992), but this does not capture the role of experts in its entirety. There is the further possibility that independent experts and their organizations take part in the enforcement of programs for firms and business groups that may participate on a voluntary basis (Power 1997). In some sectors, business is sometimes invited on a partnership basis by a group of experts so that the regulating body is composed of several actors. Whether innovative expert groups come from and represent "hard science" or whether they represent organizational expert knowledge based on insights in regulatory and administrative processes is an issue that needs to be examined more closely. Business, therefore, must always be able to match different forms of expert knowledge – an essential requirement when developing and running private schemes. Civil society organizations can undertake similar initiatives, either alone or in cooperation with experts, and there is a gray area where civil society groups and certain expert organizations intersect.

In sum, business is facing a variety of institutional and partly overlapping conditions with regard to climate change. Indeed, these conditions are likely to vary in the way they play out across different sectors that grapple with different challenges of CO_2 emissions. Thus, there are many important factors at work not only in the business community itself but also in the exchanges between business on the one hand and public institutions, civil society organizations and expert groups on the other. Adding significantly to this complexity is, of course, the current dynamics of climate policy as a new policy field seeking appropriate governance structures: some of these dynamics may facilitate private regulation, while some may impede it. Challenges do not occur exclusively in the very early stages when new private programs are in the making but, in principle, they permeate all stages of regulation. It is therefore time to discuss some of the mechanisms of designing, joining, monitoring and complying more closely.

Voluntary elements: Designing, joining, monitoring and complying

Mindful of the major institutional conditions surrounding private regulation, we can now focus on some of the elements underpinning the building and managing of programs in climate policy. There is no easy and straightforward way to analyze the potentials and pitfalls of such programs, however. We need to examine key elements of their operation and identify essential building blocks required to establish an analytical framework. "Voluntary action" is a complex term, however, and we find different conceptualizations of it (Darnall and Sides 2008; Gunningham 2007; Haufler 2001: 2–14; Webb 2004: 11–21). A basic starting point is to recognize the multidimensional character of voluntary action and to note that the performance of programs is not reducible to one single factor, but includes several interrelated elements. The stringency of programs, for instance, is an issue that relates to all the voluntary elements and cannot just be decided when designing programs, as it is also of paramount importance to know how many firms join programs and how they and their compliance are monitored.

The public-good character of private regulation in the area of climate change politics is often disputed. Whether private programs effectively contribute to halting climate change or not is a complex issue (Morgenstern and Pizer 2007), not least because there are many factors at play and because real changes do not happen in an instant. However, if private programs with the stated goal of reducing emissions are established, consolidated and joined by many firms, and if they document only a small number of infringements, then these programs can be an important tool. An additional consideration is that programs may be recognized by parties outside the business community and gain wider legitimacy.

First we take an interest in the designing of programs, because a basic criterion of success is undoubtedly the creation of an effective scheme; otherwise organized progress cannot be made. Second, we must scrutinize the problems of joining, because the emergence of a program does not imply that all relevant corporations will participate and fully commit themselves to a program. Third, we point to the monitoring capacities of the managing bodies because it is necessary to observe and verify whether firms comply, and if they do not, to develop appropriate tools to sanction cases of non-compliance. Fourth, we identify complying as a vital element, because free-riding will undermine the value of a program, leave important aspects of business behavior unregulated, or bring public institutions onto the scene. Together these elements form important evaluative criteria for assessing the performance of private voluntary programs. They are closely related in the processes of developing private voluntary programs but a simple linearity cannot be expected because moving from one element to another is a complex process and characterized by some "muddling through". Indeed, this is understandable, given the current experimental nature of private voluntary programs.

Designing

The emergence of a voluntary program is often a precarious step. Indeed, major issues must be considered, conflicts mitigated, and stable solutions found in the early stage of program and organization building (Cashore, Auld and Newsom 2004: 9–17; Darnall and Sides 2008: 96–98; Iannuzzi 2002: 155–174). The whole process of creating a scheme around new climate and emission issues is linked to complex agendas; some related standards may already be in place, or may be under consideration at the private or public level. There is the possibility that others give direct encouragement to business and decide to sponsor a program, or the reverse – that they see corporate initiatives as highly problematic and seek to avoid private initiatives. In cases when effective and recognized bodies outside the industry have emerged – with the assistance of governments, civil society groups or expert organizations – there may be no need for business to worry about concrete design problems. In fact, firms and industries may immediately jump to the issue of joining. Business may also find itself in a situation in which the other actors are rather passive, more or less ignore contributions from business, and have no profound insight into particular sectors of business. At a later stage, when programs are being implemented, they will be noticed, welcomed, or criticized.

When the necessary rules have not been formulated and there is an active interest in bringing private rules to life, it is relevant to search for inspiration. Currently, there is much experimenting with private rules and, consequently, it is relevant to draw on experiences from different firms and industries and closely study the achievements and problems of private voluntary programs – although different initiatives are summarized under this agenda (Baranzini and Thalmann 2004; Carraro and Leveque 1999; Kirton and Trebilcock 2004; Morgenstern and Pizer 2007; Webb 2004). Each firm and each industry has its own problems to grapple with, but some general features stand out and certain experiences can be applied when designing new private schemes focused on emission issues.

An overriding question is, of course, whether various standards should be designed at the level of single firms or at the level of a group of firms, most typically regarding a specific product group or an industry (Delmas and Montiel 2008; Parker 2002). Various conditions in the institutional environment may support wider rule systems and such stimuli are of great importance; but to work effectively, a coupling with individual firm strategies is useful, although different choices must be made. Firms must carefully examine whether it is possible and advantageous to establish codes and other such measures and profile themselves on an individual basis, or whether joint action is preferable. In a number of cases, such decisions must be made; but in other cases, these two options are not necessarily excludable, a point we will discuss later. At any rate, various choices must be made in the initial stage of designing private programs, and include a further number of issues – such as how to organize the regulatory body, how to finance it and how to police the behavior of business. These issues can, of course, become themes of incessant debate in and around business.

Joining

Some problems of joining are in one way or another already considered, and settled in the process of designing programs. Once established, however, firms must be mobilized and programs must be actively supported (Prakash and Potoski 2006). Joining is a classic problem in collective action theory (Olson 1965), and in our case also is a clear indication of the commitment of business. At this stage, different situations exist. If a firm decides to establish it own practices and if these are not stimulated by some pre-existing initiatives in the industry, then participating in an encompassing program is not relevant – but independent corporate initiatives may trigger a process leading to an industry-wide program. Such developments are not unlikely in relation to climate issues, because we are still in the early process of finding appropriate models of regulation.

If a general program is already available (see Cashore, Auld and Newsom 2004), firms can participate without being concerned about the initial costs of sponsoring a program and going through the demanding stages of building agreement with other firms; instead they benefit from the program already in place, just as existing members do. Participation is much easier and more straightforward when it is not coupled with prior negotiations, although there is a risk that firms not involved in the early stages of creating a program will be less committed to ownership of the program.

Other circumstances also have an impact on the process of joining. If a program, for instance, is managed by a business association, all members are typically enrolled in the program simply by virtue of their membership. Arguably, there can be a stronger attachment to the program because members will be involved in the process of creating it, assuming the program itself is not imposed on members by the management of the association. However, if participation in a program run by a business association is decoupled from membership and is voluntary, then there is flexibility in deciding whether to join.

If memberships of a program and an association are closely linked, then a firm cannot leave the program without leaving the association, and vice versa. In such contexts, participation in a program is not an isolated event, and the benefits or disadvantages are linked to a range of other choices. By leaving a program, the firm is simultaneously excluded from the association and loses the opportunity to receive selective goods, such as important information, and to influence important decisions taken by the association (Schmitter and Streeck 1981); therefore, associations can in many ways have a strong influence on the rate of participation.

The most detached and voluntary form of participation denotes situations when firms are not involved in the processes prior to the establishment of a program, either alone or through membership in another collective entity. Then choices are made with regard to the possible benefits of joining alone, and joining indicates a strong interest in the program itself rather than in some other aspects. Accordingly, firms become certified by the relevant body and follow the standard-setting of this single-purpose organization, which is orientated towards creating a regulatory framework.

These programs typically have certain admission criteria. For some gatekeeper functions, the issue of participating in a program is not determined exclusively by firms in the relevant industry but hinges on the body running the program, and this body can be largely independent of corporate interests. Completely free admission, however, would be problematic and would undercut collective efforts to raise standards.

When participation in certain programs is based on autonomous and voluntary decisions by firms, the body administering the scheme in question makes a varying degree of effort. Such bodies also engage in active recruitment. If a business association is engaged in extending its membership base, however, its program is most likely one of several benefits offered to members, whereas a body specializing in administering a program has nothing else to offer and depends on this activity. This can be an advantage in smoothing the process of participation for firms that are only motivated by the benefits of the program and are not interested in the whole package.

Monitoring

Creating programs and joining them are important activities in exploiting the potentials of private programs. They are not sufficient, however, and we have to study how programs are organized in further detail, including the central issue of monitoring. When members enrol in a program, they are expected to be highly motivated, although not always for the same reason, somewhat depending on their way in to the program – as discussed already. Notwithstanding the different origins of these motivations, they do not provide a sure guarantee that members will always comply with the standards, making shirking a key concern.

Therefore, advanced monitoring functions are essential to oversee members' behavior, and the weak implementation of effective monitoring and enforcement mechanisms is typically seen as one of the major pitfalls of private regulation (Baldwin and Cave 1999; Gunningham and Rees 1997; Haufler 2001; King and Lenox 2000). Indeed, this is one of the general arguments found against private regulation, and thus, this issue should also be of primary importance in climate policy.

But there are different ways to organize a monitoring system in the business community. One may distinguish between two basic forms of monitoring: horizontal monitoring and hierarchical monitoring (Milgrom, North and Weingast 1990). In the horizontal mode, monitoring is done through competing firms that have major incentives to police the behavior of their competitors, report infringements and assist in applying relevant forms of sanction – to ensure that standards are enforced and that firms that shirk do not undeservedly enjoy benefits such as reputational goods. Given the active participation of some firms in the process of designing programs, these members are particularly obliged to make them work.

In the hierarchical mode, monitoring is carried out through the bodies responsible for running a program. These bodies have a vital interest in member compliance, because lack of compliance will not only damage individual firms in the program, but have repercussions on the regulatory arrangement as a whole. Success is very much related to the number of certified firms – and also to compliant behavior, of course. Therefore, regulatory bodies may independently police the behavior of members, punish non-compliance and ultimately expel members from these kinds of clubs (Buchanan 1965) – although such radical sanctions are not often used (Darnall and Sides 2008: 111). It can be essential to establish rigorous monitoring to deter members from shirking, and thereby make a program credible in the business community and beyond. Thus, it is important that not only public institutions but also business and standard-setting organizations host various enforcement mechanisms (Braithwaite 2002; Prakash and Potoski 2006: 57–62).

While these two mechanisms of monitoring are quite basic, their application may differ. Members are in many ways socialized into the work of associations to which they have a variety of ties, certain patterns of behavior are ingrained, and, therefore, they feel a strong commitment to self-report and assist the organization. Most decisive is probably the fact that firms feel some kind of ownership of the program itself.

Although we cannot establish beforehand which mechanisms are most widely applied and best suited to tackle climate change and emission problems, we suggest that both forms of monitoring can work, alone or in tandem, because monitoring capacities are scattered in the world of private regulation. At a later stage, when experiences are gathered in relation to monitoring emission-related programs, some models will perhaps emerge, but it is still premature to identify specific mechanisms.

In addition to these mechanisms, of course, are the monitoring powers held beyond business and involving those public and private actors conditioning private programs and making them more legitimate (Bernstein and Cashore 2007; Black 2008). Therefore, a much more varied picture of monitoring emerges. Taken together, these different resources can in many ways consolidate the implementation of private programs, but as always, there is a possibility that public regulation will be considered at some point, should there be serious problems with private regulation.

Complying

Monitoring is a tool to prevent free-riding by those subscribing to a program. If it is first in place and members are familiar with the effectiveness of horizontal and hierarchical ways of monitoring, there are reasons to believe that deterrence makes compliance much stronger (Coleman 1994; Iannuzzi 2002: 4–12). Still, good monitoring is not always enough. For various reasons, infringements occur: firms shirk, are not familiar with all aspects of the rules, and have difficulties interpreting and applying new and complex rules. Although some infringements can be productive in the sense that they raise important questions as to the appropriateness of rules, and may thus lead to a reorganization of the rule system and its administration, the lack of compliance basically undermines a given program.

Lack of compliance may have different implications. In cases of non-compliance with programs run by associations (Ogus 1995), there is the risk that members will be expelled from the association. This is a decision that will do immediate harm to the member in question but perhaps also to the association which, as a consequence, will become less representative of the industry concerned and will have fewer opportunities to participate in fruitful dialog with the alienated non-compliant and bring it better in line with associational strategies. If a firm is a member of an isolated program and not linked to other organizations, activities must be seen in a different light as the same contextual factors are not then at play. Here, however, programs may also suffer from non-compliance, and private regulation will appear to be a less relevant alternative to public regulation in climate politics. This requires that a variety of sanctioning mechanisms be brought into use, the basic and most serious, of course, being to withdraw certification and expel a firm from a program.

To achieve compliance, it is furthermore necessary to have some mechanisms to disseminate important information on the working of programs, and to make this information available to the general public (Greif, Milgrom and Weingast 1994; Gunningham 1991). In fact, different segments of the public have different requirements for information, and firms involved in a specific sector are inclined to seek other kinds of information than the general public; the informational problem is one of the classic issues in seeking compliance and making regulation work (Fisse and Braithwaite 1983).

Indeed, different sorts of reports can contain useful information about major achievements but also point to various flaws in existing programs.

Reporting mechanisms can provide interested parties with key information and thus play a vital role in making markets work efficiently. These measures, in turn, can be used by civil society and expert organizations to professionalize a dialog and improve existing practices, and sometimes they include the formal involvement of these groups (Yaziji and Doh 2009). The development of a relevant information strategy at the level of single firms and collective entities is therefore an element that must be given due attention in the analysis of program performance. An inherent dilemma, of course, is that it can work both ways, but we see a proper information strategy as an element that will ultimately benefit a scheme rather than erode it.

There are several major elements of voluntariness in the process of establishing private programs. It is useful to distinguish the four elements of designing, joining, monitoring and complying, which are related to different stages of program building, thus recognizing that the commitment of firms may vary across the process. Voluntary programs do not always work instantaneously but are often shaped in an evolutionary process in which it takes time for programs to mature and become accepted in the business community as well as by society at large, factors we also must account for in our study of climate politics. Certain programs are born with weaknesses that are hard to overcome, and these may persist even beyond their infancy stage. Joining, monitoring and complying are all important in their own right and deserve special scrutiny, and for programs to be effective they must attend to all these elements, although it would be strange if all were managed equally successfully. Indeed, each of the four elements outlined above may contain pitfalls or hold potentials. It is essential that programs take off, but even at the design stage, some will fail. When programs are actually introduced, however, they will not reach their full capacity and be able to solve all problems from day one; in addition, performance is likely to vary across industries.

Cross-sectoral, multisectoral and sectoral experiences

There are a host of different approaches to the study of climate policy. We can, for instance, start from a territorial perspective and scrutinize particular countries or regions; we can adopt a resource perspective and examine various types of renewable and non-renewable resources; we can investigate the world's different ecosystems and investigate how they are affected by climate change, and so on. In this book we analyze climate policy from the perspective of private regulation with a particular emphasis on the role of business.

Private voluntary programs focus attention on corporate behavior, and the active role of the organizations behind the schemes. There can be significant variations across business, however, and therefore we study three major types of arrangements currently in place in climate policy: cross-sectoral programs whose norms and rules may have implications for the entire business community and whose members represent a horizontal structure in business, such as the financial industries; multisectoral programs whose norms and rules may have consequences for business more generally and which to varying degrees are open to firms from different industries, such as waste and energy saving; and sectoral programs whose norms and rules pertain to specific industries and which are only open to firms from specific industries, such as food, forestry or automobiles. It goes without saying that this book is not able to embrace every area of corporate activity, but the different organizational formats expounded are illustrative of the challenges confronting wide parts of the business community.

The challenge of climate change is not reducible to a particular industry and its dynamics. One common issue, however, relates to the use of energy in the form of fossil fuels, which is responsible for a significant part of CO_2 emissions. It is important to remember that the energy industry constitutes a horizontal structure in business – in a number of ways similar to the financial industries – encompassing several fossil fuels as well as the new field of renewables.

It is critical that the energy industry finds ways to use traditional energy more efficiently and develop new and alternative sources; however, it is equally important to bear in mind that innovative answers are not always provided in the horizontal structures of business and then introduced later in all other sectors just idly waiting for relevant solutions. Writing on the industrial revolution, David Landes once stressed how changes were interlinked across industries, and that "many technical improvements were feasible only after advances in associated fields" (Landes 1969: 2). The same holds true for the contemporary industrial adaptations to the challenges of climate change: changes are interdependent and, hence, experience must be won and learning be derived from different areas of industrial activity in which private voluntary programs materialize. Changes are accomplished within and across many areas and are not governed by any single industry or any single agenda; therefore it is imperative to scrutinize creative experimentation in different industries to achieve a more coherent understanding of private voluntary programs and their potentials and pitfalls.

In-depth studies of experiments within cross-sectoral, multisectoral and sectoral programs acknowledge that industries face different challenges in reducing emissions because of the different patterns of production and consumption underlying these sectors, and also because of the variation in regulatory traditions. These opportunities for building and running private programs are likely to exert influence on the character and performance of programs and the voluntary elements of designing, joining, monitoring and complying.

We begin with the chapters on cross-sectoral and multisectoral programs. The financial industries as a cross-sectoral structure have been undergoing an interesting evolution in relation to climate change as a variety of new entities have emerged and face up to existing assumptions and traditions in this specific sector. Then follows a study on multisectoral programs. In some respects such programs lack a clear identity but the rather broad appeal of such programs can make them attractive to a wide group of firms. Indeed, changes in encompassing cross-sectoral and multisectoral fields have a great potential to stimulate transformations elsewhere, but incentives have to be picked up in individual industries or by individual firms.

In the chapters on sectoral programs we analyze some important industries, each of which grapples with problems related to the emission of greenhouse gases, and each of which is located at different points down the production chain. Forestry represents an ancient sector and often the natural world in its true sense, with different patterns of production in North and South, currently reducing the opportunities for using forests as sinks, which are critical to solving climate problems; food is an encompassing sector that includes a wide range of agricultural products, which today, however, are often heavily industrialized and link different parts of the globe by virtue of mass consumption; cars are a sector again related to other industries and heavily dependent on fossil fuels, are essential for personal transport on a very large scale and are so closely embedded in modern lifestyles that societies are hardly imaginable without them. We take a principal interest in identifying the potentials and pitfalls of the cross-sectoral, multisectoral and sectoral programs, whether weaknesses and strengths can be traced in relation to designing, joining, monitoring and complying, and how institutional conditions have implications for the emergence and performance of programs.

All contributions investigate these central questions by applying and further developing the general analytical framework laid out in this introductory chapter and, consequently, different research priorities are set. Certain conflicts between the public goals of climate policy and the private approaches are given special emphasis and to varying degrees inspiration is drawn from theories on the rent-seeking behavior of firms and from theories on the constraints facing corporate action. Furthermore, certain of the core elements of program-building are in some chapters examined in greater detail simply because they have been especially critical for programs in a particular area and, therefore, deserve closer attention. Finally, certain types of relationships are explored more than others simply because some relations are more relevant and point to various coalitions of public and private actors. Such priorities, however, do not exclude but go hand in hand with a broader analysis of the framework in which programs are embedded.

In Chapter 2 by Jennifer Clapp and Jason Thistlethwaite – "Private Voluntary Programs in Environmental Governance: Climate Change and the Financial Sector" – voluntary corporate initiatives in the financial sector and their rapid expansion in both number and scope over the past decade are studied. At present, firms can adopt and participate in a wide range of initiatives. Together with the analytical framework presented in Chapter 1, Clapp and Thistlethwaite draw on inspirations from experiences with private regulation to develop a perspective that is particularly suited to explore programs in the financial sector. Although aspirations are high for these voluntary corporate initiatives, this chapter examines their potentials and flaws as a strategy in the pursuit of sustainable development. It argues that, while such measures are a popular and important part of the overall strategy for addressing pressing environmental problems such as climate change, reliance on such measures alone would be problematic; various barriers are identified. The chapter examines the reasons for the weakness of corporate initiatives as stand-alone measures, which include a lack of incentive to go beyond cost-saving measures, accountability deficits, malfunctioning voluntary environmental markets, and the simple fact that while some leading firms join and even champion such measures, many firms do not participate. The chapter illustrates these problems in particular in the area of climate change, with a closer look at the types of measures taken by firms as part of their individual and collective commitments, including their participation in voluntary carbon markets.

Chapter 3 by Lily Hsueh and Aseem Prakash is entitled "Private Voluntary Programs on Climate Change: US Federal Government as the Sponsoring Actor". The authors focus on the proliferation of multiple private voluntary programs over the past two decades which have been sponsored by the federal and state governments in the US, as well as by civil society groups and trade associations. Currently, thousands of firms and organizations participate in programs across a large number of industries. Given the inaction of the Bush Administration on the issue of climate change, voluntary programs have emerged as an important policy tool in the domestic context, but also have a role to play in global problem-solving. The chapter surveys private voluntary programs on climate change and their various features in sectors that are the key emitters of greenhouse gases. Their analytical objective is to examine these programs through club theory. Voluntary clubs reduce information asymmetries between external stakeholders and firms regarding the internal policies that firms have adopted. In their case, clubs seek to signal participating firms' credible commitment to reduce the emission of greenhouse gases. In return for incurring such costs, club members hope to reap excludable reputational benefits that accrue to members. However, the club approach recognizes that clubs vary in their capacities to generate positive externalities, because clubs differ on how successful joining is and on how strong compliance is. In conclusion, lessons for global climate policy, voluntary programs and club theory are identified, and the promises and pitfalls of clubs are discussed.

Turning to specific industries, in Chapter 4 Laura Bozzi, Benjamin Cashore, Kelly Levin and Constance McDermott examine "The Role of Private Voluntary Climate Programs Affecting Forests: Assessing their Direct and Intersecting Effects". They hold that any effort to understand and assess the role of private governance and climate must pay careful attention to its current and potential impact on global forest management on the ground. Forests are both significant sources and significant sinks of carbon, with variations across boreal, temperate and tropical forests. The extent to which forest management is contributing to climate change and providing mitigating solutions is, however, hotly contested among business interest associations and conservation-focused environmental groups, as well as among states. As to the impacts of climate-related private regulation initiatives for the global forest sector, two trends are discernable. First, private voluntary programs have been proposed as a means to ensure that public carbon-related forest policy efforts, such as those generated through the "Clean Development Mechanism" (CDM) and reduced emissions from deforestation and degradation (REDD) initiatives, do not negatively impact non-climate environmental and social conditions. Indeed, private certification of responsible forest management may become part of a broader mitigation strategy. Second, the role of purely private mechanisms, that defray forest exploitation in order to offset the CO_2 impact of environmentally aware consumers, must be related to the efforts of public institutions, and the degree to which private and public initiatives interact. The authors assess prospects for support of different private and public pathways, and the effects that they might have, if fully implemented.

Chapter 5 by Doris Fuchs and Frederike Boll – "Emerging Private Voluntary Programs and Climate Change: The Blind Spots of the Agrifood Sector" – notes that numerous processes in the agrifood sector impact on climate change. Simultaneously, quite a number of private initiatives in the agrifood sector exist, at the level of both single corporations and collective entities. They ask to what extent these initiatives address the issue of climate change, and how effective they are. And what are the determinants of the extent and effectiveness of private initiatives? The chapter answers these questions by analyzing a range of relevant private initiatives in the agrifood sector, and the authors develop some relevant theoretical tools to understand the behavior of firms. The analysis shows that the vast majority of initiatives neglect the issue of climate change almost completely, and that those initiatives that address climate change aspects tend to promise little effectiveness. The chapter argues that the reasons for the neglect and lack of effectiveness are threefold. First, the link between the agrifood sector and climate change is not immediately visible to the public (the consumer), and therefore there is a corresponding lack of pressure on the actors creating and implementing the respective private institutions to integrate climate change objectives. Second, the aspects and processes of the agrifood sector affecting climate change relate to such a breadth of activities and actors that they provide a difficult target for civil society actors. Third, addressing the most important sources of climate change in the agrifood sector would imply fundamental changes in the sector's overall design and functioning rather than the regulation and optimization of certain specific processes.

Tony Porter analyses cars in Chapter 6 on "Climate Change, Private Voluntary Programs and the Automobile Industry". The transportation sector is estimated to contribute 30 per cent of the CO_2 emissions of the OECD countries and 20 per cent worldwide, and these figures are likely to rise. Thus, controlling auto emissions is a crucial part of addressing the problem of climate change. Seeking inspiration from governance theories, Porter asks what role is played, and should be played, by private voluntary programs and other forms of private regulation in addressing the problem of climate change, including controlling auto emissions. A key point is that these private programs are embedded in broader contexts, and he explores how governments can in many ways either constrain or further private regulation. The optimal mix of public and private actors and rules in addressing such challenges, including which tasks should be allocated across public and private institutions, is a question that has been dealt with in the literature but not adequately answered. One reason that traditional forms of public regulation interact with private voluntary programs that have been developed in recent years is the enormous complexity of contemporary societies. Automobile manufacturing is transnational, dominated by a relatively few assemblers, but assemblers increasingly rely on transnational networks of parts manufacturers, some of them small, and others huge transnational firms. The complexity of private regulation is further exacerbated by the fact that the car industry intersects with other relatively autonomous fields of activity, including the oil industry and the built environment of cities and roads.

Chapter 7, the concluding chapter of the book, discusses the many experiences that have been gained with private voluntary programs in climate policy. It is a new and emerging field of regulation, but we already have some indications on when programs work and when they do not, and how climate policy can be married with private voluntary programs. Important factors are the different material positions of the industries, as they have very different options for reducing emissions, as well as the core economic motives of firms. However, a range of factors also count because public as well as private actors appear to be important stakeholders in relation to the programs adopted. We further learn that the performance of programs is not a simple process but is best studied through elements such as designing, joining, monitoring and complying. Indeed, for private regulation to succeed in climate policy and become a viable alternative or supplement to public regulation, several parameters must be accounted for. These experiences can help us identify where private voluntary programs are inappropriate mechanisms for mitigating climate change, but they can also inform and improve existing programs and be helpful in extending this regulation to new areas of business activity.

Notes

1. A further number of agencies not pertaining to any specific industry cover aspects of climate policy. This applies, for instance, to the International Organization for Standardization (ISO) which adopted a new voluntary standard on greenhouse gases (ISO 14064) in 2006.
2. An alternative approach would be to see voluntary private programs systematically through the lens of policy processes (Porter and Ronit 2006). In this perspective self-regulation bears many of the same features as traditional policy processes (agenda-setting, decision, implementation and evaluation). In this book, however, we begin the analysis at a stage where programs are already being designed and keep agenda-setting somewhat in the background. Furthermore, programs have in general not been through much evaluation, and, consequently, it would be difficult to treat evaluation in a thorough manner. In sum, we denote designing, joining, monitoring and complying as relevant elements when analyzing the new experiences with private voluntary programs in climate policy.
3. Different literatures analyze business collective behavior. In the generic club literature scant attention is drawn to business interest associations, while the special literature on business interest associations only addresses certain aspects of club theory. A better integration of these strands of research is needed.
4. As a civil society initiative CAN has achieved an impressive degree of coordination in which well-known organizations, such as Greenpeace and WWF, are also included.
5. This organization is now named the Center for Climate and Energy Solutions (C2ES).
6. We cannot talk about a single and unified expert community, however. Today there is wide agreement about the importance of the human factor in climate change, but the scientific community is not a unitary actor. Not only are different disciplines represented, but different data and estimates are also offered, giving rise to uncertainties and disputes.

REFERENCES

Aldy, J. E., and R. N. Stavins, eds. 2010. *Post-Kyoto International Climate Policy. Implementing. Architectures for Agreement.* Cambridge: Cambridge University Press.

Axelrod, R. S., D. L. Downie, and N. J. Vig, eds. 2005. *The Global Environment: Institutions, Law and Policy*. Washington DC: CQ Press.

Baldwin, R., and M. Cave. 1999. *Understanding Regulation. Theory, Strategy and Practice*. Oxford: Oxford University Press.

Baranzini, A., and P. Thalmann, eds. 2004. *Voluntary Approaches in Climate Policy*. Cheltenham: Edward Elgar.

Barber, J. 1998. "Responsible Action or Public Relations? NGO Perspectives on Voluntary Initiatives." *Industry and Environment* 21 (1–2): 19–22.

Bendell, J. ed. 2000. *Terms for Endearment. Business, NGOs and Sustainable Development*. Sheffield UK: Greenleaf Publishing.

Bernstein, S., and B. Cashore. 2007. "Can Non-State Global Governance be Legitimate? An Analytical Framework." *Regulation & Governance* 1 (4): 347–371.

Biermann, F., and S. Bauer. 2005. *A World Environment Organization*. Aldershot: Ashgate.

Biermann, F., P. Pattberg, H. van Asselt, and F. Zelli. 2009. "The Fragmentation of Global Governance Architectures: A Framework for Analysis." *Global Environmental Policy* 9 (4): 14–40.

Biermann, F., F. Zelli, P. Pattberg, and H. van Asselt. 2010. "The Consequences of a Fragmented Climate Governance Architecture: A Policy Appraisal." In *Global Climate Governance Beyond 2012: Architecture, Agency, and Adaptation*, eds. F. Biermann, P. Pattberg, and F. Zelli, 24–34. Cambridge: Cambridge University Press.

Black, J. 2008. "Constructing and Contesting Accountability and Legitimacy in Polycentric Regulatory Regimes." *Regulation & Governance* 2: 137–164.

Boddewyn, J. J. 1985. "Advertising Self-Regulation: Organization Structures in Belgium, Canada, France and the United Kingdom." In *Private Interest Government: Beyond Market and State*, eds. W. Streeck and P. C. Schmitter, 30–43. London: Sage.

Braithwaite, J. 2002. "Rewards and Regulation." In *New Directions in Regulatory Theory*, eds. Sol Piciotto and David Campbell, 12–26. Oxford: Blackwell.

Buchanan, J. 1965. "An Economic Theory of Clubs." *Economica* 32: 1–14.

Bulkeley, H., and P. Newell. 2010. *Governing Climate Change*. London and New York: Routledge.

Carpenter, C. 2001. "Businesses, Green Groups and the Media. The Role of Non-Governmental Organizations in the Climate Change Debate." *International Affairs* 77 (2): 313–328.

Carraro, C., and F. Leveque, eds. 1999. *Voluntary Approaches in Environmental Policy.* Dordrecht: Kluwer.

Cashore, B., G. Auld, and D. Newsom. 2004. *Governing through Markets. Forest Certification and the Emergence of Non-State Authority*. New Haven CT: Yale University Press.

Chambers, W. B., and J. F. Green. 2005. *Reforming International Environmental Governance: From Institutional Limits to Innovative Reforms*. Tokyo: United Nations University Press.

Clapp, J. 1998. "The Privatization of Global Environmental Governance: ISO 14000 and the Developing World." *Global Governance* 4: 295–316.

Coleman, W. D. 1994. "Governing the Securities Industry in Canada, the United Kingdom and the United States." In *Governing Capitalist Economies: Performance and Control of Economic Sectors*, eds. J. R. Hollingsworth, P. C. Schmitter, and W. Streeck, 244–269. Oxford: Oxford University Press.

Coleman, W. D., and T. Porter. 2000. "International Institutions, Globalization and Democracy: Assessing the Challenges." *Global Society* 14 (3): 377–398.

Conzelmann, T., and K. D. Wolf. 2008. "The Potential and Limits of Governance by Private Codes of Conduct". In *Transnational Private Governance and its Limits*, eds. J. C. Graz and A. Nölke, 98–114. London: Routledge.

Cornes, R., and T. Sandler. 1986. *The Theory of Externalities, Public Goods, and Club Goods*. Cambridge: Cambridge University Press.

Crane, A., D. Matten, and J. Moon. 2008. *Corporations and Citizenship*. Cambridge: Cambridge University Press.

Cutler, A. C., V. Haufler, and T. Porter. 1999. "Private Authority and International Affairs". In *Private Authority and International Affairs*, eds. A. C. Cutler, V. Haufler, and T. Porter, 3–28. Albany: State University of New York Press.

Darnall, N., and S. Sides. 2008. "Assessing the Performance of Voluntary Environmental Programs: Does Certification Matter?" *Policy Studies Journal* 36 (1): 95–117.

Delmas, M., and I. Montiel. 2008. "The Diffusion of Voluntary International Management Standards: Responsible Care, ISO 9000, and ISO 14001 in the Chemical Industry." *Policy Studies Journal* 36 (1): 65–93.

DeSombre, E. R. 2006. *Global Environmental Institutions.* London and New York: Routledge.

Dessler, A. E., and E. A. Parson. 2006. *The Science and Politics of Global Climate Change: A Guide to the Debate*. Cambridge: Cambridge University Press.

Doh, J. P., and H. Teegen, eds. 2003. *Globalization and NGOs. Transforming Business, Government and Society*. Westport CT: Praeger.

Fisse, B., and J. Braithwaite. 1983. *The Impact of Publicity on Corporate Offenders.* Buffalo: State University of New York Press.

Fuchs, D., and Vogelmann, J. 2008. "The Power of TNCs in Transnational Environmental Private Governance." In *Transnational Private Governance and its Limits*, eds. J. C. Graz and A. Nölke, 71–83. London: Routledge.

Graz, J. C., and A. Nölke. 2008. "Introduction: Beyond the Fragmented Debate on Transnational Private Governance." In *Transnational Private Governance and its Limits*, eds. J. C. Graz and A. Nölke, 1–26. London: Routledge.

Greif, A. P., P. Milgrom, and B. Weingast. 1994. "Coordination, Commitment, and Enforcement. The Case of the Merchant Guild." *Journal of Political Economy* 102 (4): 745–776.

Gunningham, N. 1991. "Private Ordering, Self-Regulation, and Futures Markets: A Comparative Study of Informal Control." *Law & Policy* 19 (4): 297–326.

Gunningham, N. 2007. "Reconfiguring Environmental Regulation: Next-Generation Policy Instruments." In *Industrial Innovation and Environmental Regulation: Developing Workable Solutions*, eds. S. Parto, and B. Herbert-Copley, 200–232. Tokyo: United Nations University Press.

Gunningham, N., and J. Rees. 1997. "Industry Self-Regulation: An Institutional Perspective." *Law & Policy* 19: 365–413.

Gunningham, N., and D. Sinclair. 2002. *Leaders and Laggards: Next Generation Environment Regulation*. Sheffield, UK: Greenleaf Publishing.

Gunningham, N., R. A. Kagan, and D. Thornton. 2003. *Shades of Green: Business, Regulation, and Environment.* Stanford CA: Stanford University Press.

Haas, P. M. 1992. "Introduction: Epistemic Communities and International Policy Coordination." *International Organization* 46: 1–36.

Harrison, K., and L. M. Sundstrom, eds. 2010. *Global Commons, Domestic Decision: The Comparative Politics of Climate Change*. Cambridge and London: The MIT Press.

Haufler, V. 2001. *A Public Role for the Private Sector: Industry Self-Regulation in a Global Economy.* Washington DC: Carnegie Endowment for International Peace.

Hemmati, M., F. Dodds, J. Enyati, and J. McHarry. 2002. *Multi-Stakeholder Processes for Governance and Sustainability: Beyond Deadlock and Conflict.* London: Earthscan.

Hollingsworth, J. R., P. C. Schmitter, and W. Streeck, eds. 1994. *Governing Capitalist Economies: Performance and Control of Economic Sectors.* Oxford: Oxford University Press.

Iannuzzi, A. Jr. 2002. *Industry Self-Regulation and Voluntary Environmental Compliance*. Boca Raton FL: Lewis Publishers.

IPCC (2007). *IPCC 2007. Climate Change 2007: Mitigation. Contribution of Working Group III to the fourth assessment report of IPCC.* Metz, B., O. R. Davidson, P. R. Bosch, R. Dave, and L. A. Meyer, eds., Cambridge: Cambridge University Press.

Kaul, I., and K. Le Goulven. 2003. "Institutional Options for Producing Global Public Goods." In *Providing Global Public Goods: Managing Globalization*, eds. I. Kaul, P. Conceição, K. Le Goulven, and R. U. Mendoza, 371–409. Oxford: Oxford University Press.

Kaul, I., I. Grunberg, and M. Stern. 1999. *Global Public Goods: International Cooperation in the 21st Century*. Oxford: Oxford University Press.

Keohane, R. O., and E. Ostrom. 1995. "Introduction". In *Local Commons and Global Interdependence: Heterogeneity and Cooperation in Two Domains*, eds. R. O. Keohane and E. Ostrom, 1–26. Thousand Oaks CA: Sage.

King, A. A., and M. J. Lenox. 2000. "Industry Self-Regulation without Sanctions: The Chemical Industry's Responsible Care Program." *Academy of Management Journal* 43: 698–716.

Kirton, J. J., and M. J. Trebilcock, eds. 2004. "Introduction. Hard Choices and Soft Law in Sustainable Global Governance." In *Hard Choices, Soft Law: Voluntary Standards in Global Trade, Environment and Social Governance*, eds. J. J. Kirton and M. J. Trebilcock, 3–29. Aldershot, UK: Ashgate.

Landes, D. S. 1969. *The Unbound Prometheus: Technological Change and Industrial Development in Western Europe from 1750 to the Present.* Cambridge: Cambridge University Press.

Levy, D. L. 2005. "Business and the Evolution of the Climate Regime: The Dynamics of Corporate Strategies." In *The Business of Global Environmental Governance*, eds. D. L. Levy and P. J. Newell, 73–104. Cambridge MA: MIT Press.

Lipschutz, R. D. with J. K. Rowe. 2005. *Globalization, Governmentality and Global Politics. Regulation for the Rest of Us*. London and New York: Routledge.

Luterbacher, U., and D. F. Sprinz, eds. 2001. *International Relations and Global Climate Change.* Cambridge MA: MIT Press.

Miles, E. L., A. Underdal, S. Andresen, J. Wettestad, and J. B. Skjaeseth. 2001. *Environmental Regime Effectiveness: Confronting Theory with Evidence.* Cambridge MA: MIT Press.

Milgrom, P. R., D. C. North, and B. R. Weingast. 1990. "The Role of Institutions in the Revival of Trade: The Law Merchant, Private Judges, and the Champagne Fairs." *Economics and Politics* 2: 1–23.

Morgenstern, R. D., and W. A. Pizer, eds. 2007. *Reality Check: The Nature and Performance of Voluntary Environmental Programs in the United States, Europe and Japan.* Washington DC: Resources for the Future Press.

Newell, P. 2000. *Climate for Change: Non-State Actors and the Global Politics of the Greenhouse.* Cambridge: Cambridge University Press.

Newell, P. J. 2005. "Business and International Environmental Governance: The State of the Art." In *The Business of Global Environmental Governance*, eds. D. L. Levy and P. J. Newell, 21–45. Cambridge MA: MIT Press.

Newell, P. and M. Paterson. 2010. *Climate Capitalism: Global Warming and the Transformation of the Global Economy.* Cambridge: Cambridge University Press.

Ogus, A. 1995. "Rethinking Self-Regulation." *Oxford Journal of Legal Studies* 15 (1): 97–108.

Olson, M. 1965. *The Logic of Collective Action: Public Goods and the Theory of Goods*. Cambridge MA: Harvard University Press.

Orr, S. K. 2006. "The Evolution of Climate Policy – Business and Environmental Organizations between Alliance Building and Entrenchment." In *Global Public Policy: Business and the Countervailing Powers of Civil Society*, ed. K. Ronit, 154–173. New York: Routledge.

Parker, C. 2002. *The Open Corporation: Effective Self-Regulation and Democracy.* Cambridge: Cambridge University Press.

Pattberg, P. H. 2007. *Private Institutions and Global Governance: The New Politics of Environmental Sustainability*. Cheltenham, UK: Edward Elgar.

Pettenger, M., ed. 2007. *The Social Construction of Climate Change: Power, Knowledge, Norms, Discourses.* London: Ashgate.

Pinkse, J., and A. Kolk. 2009. *International Business and Global Climate Change.* London and New York: Routledge.

Porter, M., and C. van der Linde. 1995. "Green and Competitive: Ending the Stalemate." *Harvard Business Review* September–October: 120–134.

Porter, T., and K. Ronit. 2006. "Self-regulation as Policy Process: The Multiple and Criss-Crossing Stages of Private Rule-Making." *Policy Sciences* 39: 41–72.

Power, M. 1997. *The Audit Society: Rituals of Verification*. Oxford: Oxford University Press.

Prakash, A. 2000. *Greening the Firm*. Cambridge: Cambridge University Press.

Prakash, A., and M. Potoski. 2006. *The Voluntary Environmentalists: Green Clubs, ISO 14001, and Voluntary Environmental Regulation*. Cambridge: Cambridge University Press.

Raustiala, K. 2001. "Nonstate Actors in the Global Climate Regime." In *International Relations and Global Climate Change*, eds. U. Luterbacher and D. F. Sprinz, 95–117. Cambridge MA: MIT Press.

Reinicke, W. H. 1998. *Global Public Policy*. Washington DC: Brookings Institution Press.

Rondinelli, D., and T. London. 2003. "How Corporations and Environmental Groups Cooperate: Assessing Cross-Sector Alliances and Collaborations." *Academy of Management Executive* 17 (1): 61–76.

Ronit, K. 2007. "Introduction: The New Policy Arrangements of Business and Countervailing Groups." In *Global Public Policy: Business and the Countervailing Powers of Civil Society*, ed. K. Ronit, 1–41. London and New York: Routledge.

Ronit, K., and V. Schneider. 1999. "Global Governance through Private Organizations." *Governance* 12 (3): 243–266.

Ronit, K., and V. Schneider. 2000. "Private Organizations and their Contribution to Problem-Solving in the Global Arena." In *Private Organizations in Global Politics*, eds. K. Ronit and V. Schneider, 1–33. London and New York: Routledge.

Schmidheiny, S. 1992. *Changing Course: A Global Business Perspective on Development and the Environment*. Cambridge MA: MIT Press.

Schmitter, P. C., and W. Streeck. 1981. *The Organization of Business Interests: A Research Design to Study the Associative Action of Business in the Advanced Industrial Societies of Western Europe*. Berlin: Wissenschaftszentrum.

Schroeder, H., L. A. King, and S. Tay. 2008. "Contributing to the Science–Policy Interface: Policy Relevance of Findings on the Institutional Dimensions of Global Environmental Change." In *Institutions and Environmental Change: Principal Findings, Applications, and Research Frontiers*, eds. O. Young, L. A. King, and H. Schroeder, 261–275. Cambridge MA: MIT Press.

Smith, J. 2008. *Social Movement for Global Democracy*. Baltimore MD: Johns Hopkins University Press.

Stigler, G. J. 1971. "The Theory of Economic Regulation." *Bell Journal of Economics and Management Science* 2 (1): 3–21.

Streeck, W., and P. C. Schmitter. 1985. "Community, Market, State – and Associations? The Prospective Contribution of Interest Governance to Social Order." In *Private Interest Government: Beyond Market and State*, eds. W. Streeck and P. C. Schmitter, 1–29. London: Sage.

Victor, D. G. 2001. *The Collapse of the Kyoto Protocol and the Struggle to Slow Global Warming*. Princeton NJ: Princeton University Press.

Vogel, D. (2005). *The Market for Virtue: The Potential and Limits of Corporate Social Responsibility.* Washington DC: Brookings Institution Press.

Webb, K., ed. 2004. *Voluntary Codes, Private Governance, the Public Interest and Innovation*. Ottawa: Carleton Research Unit for Innovation, Science and Environment.

Yaiji, M., and J. Doh. 2009. *NGOs and Corporations: Conflict and Cooperation.* Cambridge: Cambridge University Press.

Part II

Cross-sectoral and multisectoral programs

2

Private voluntary programs in environmental governance: Climate change and the financial sector

Jennifer Clapp and Jason Thistlethwaite

Introduction

Private voluntary programs for environmental improvement expanded rapidly in both number and scope over the past decade, particularly at the international level. At present, a wide range of international "green" initiatives is available to firms, including those specifically geared toward mitigating climate change. Voluntary climate programs range from those targeted at specific production sectors such as automobiles, food or forests, dealt with later in this book, to those that cut across sectors, such as measures that embrace the financial industry and its practices. Because financial firms are the gatekeepers to capital for the real economy, they represent leverage points that have significant influence across the economy as a whole. Voluntary private programs such as these are increasingly promoted by states as viable means by which to promote sustainable development. Despite this attention, it is still unclear whether participation in private voluntary programs results in significant improvements in the environmental performance of firms.

In this chapter we focus on problems identified with private voluntary programs in general, and assess whether those problems play themselves out in voluntary climate change programs designed for the financial sector. We argue that although aspirations are high for international voluntary corporate initiatives broadly, these initiatives embody flaws that must be taken seriously. For this reason, we argue, any suggestion that state-based regulation should be scaled back for firms engaging in

Business and climate policy: The potentials and pitfalls of private voluntary programs, Ronit (ed.), United Nations University Press, 2012, ISBN 978-92-808-1214-5

voluntary measures should be viewed skeptically. In the case of private voluntary programs that embrace the financial sector, the problems tend to be amplified because many climate-based voluntary measures are connected to one another. In this way, these measures may limit the financial sector's ability to impose a cross-sectoral cost structure on greenhouse gas (GHG) emissions.

In the first part of the chapter, we present a general typology of voluntary private environmental programs at the international level, building on the framework from Chapter 1. These include broad programs such as principles or codes of conduct, reporting mechanisms, environmental management systems and voluntary environmental markets. We then map out some of the key weaknesses of these different types of voluntary programs which result in shortfalls in terms of both environmental effectiveness and corporate accountability. The second part of the chapter examines climate-based voluntary programs that incorporate the financial sector and its practices.

This analysis addresses an important gap within academic literature. While some scholars within global environmental politics have analyzed the link between financial markets and climate change, they have yet to pay specific attention to the effectiveness of the emerging private voluntary programs that define this link (see Jagers, Paterson and Stripple 2004; Paterson 2001). It is important to rectify this oversight for two reasons. First, unlike some private programs where firms are pressured to join, based on their poor reputation in the area of environmental management, initiatives in the financial industry are often designed around a business case for governing exposure to climate change risks. These risks primarily include physical risks linked with impact of extreme weather on financial returns and regulatory risks linked with the impact of new regulatory costs. In fact, many of the programs analyzed later in the chapter are designed to develop market or "best" practices that the financial industry can use to reduce its exposure to these risks. The second and related reason is that these climate change risks are generated outside the financial industry among clients using financial services, rather than among financial firms themselves. For this reason, private voluntary programs constitute an important opportunity to leverage the influence of this industry over the firms they invest in to encourage behavior that mitigates exposure to these risks, such as the emission of GHGs. Our analysis constitutes the first attempt to evaluate the effectiveness of these programs in achieving their ambitious objectives.

There are different schemes in the financial sector and they have been shaped under different institutional conditions, some emerging in business, some in cooperation with other private stakeholders, and some created in the public sector. Thus, several parties are involved in designing and operating them. Despite the diversity of schemes and actors involved, most of these initiatives share a common "monitoring" governance approach. In terms of organization, most of the initiatives are designed using what Chapter 1 identifies as a "hierarchical monitoring" where an external actor collaborates with financial firms in monitoring compliance to the program. The long-term goal of these initiatives, however, is to create "horizontal monitoring" where financial firms, and the businesses they invest in, self-regulate or compete with each other in meeting standards necessary to reduce their exposure to climate change risks. Using our general typology, we examine a number of climate-related voluntary programs involving this sector, and argue that they are linked to one another in important ways. In particular, these programs act as mechanisms to account and mitigate GHG emissions in ways that inform the financial sector on how to allocate its vast capital resources to create incentives that reduce worldwide GHG emissions. More specifically, these initiatives are designed to generate new rules and norms within the financial industry that link both a firm's contribution and exposure to climate change to a financial risk that must be governed to improve capital allocation decision-making. But, as our analysis will show, the various linkages and overlaps between the programs compound the weaknesses that are present in each of these various types of voluntary initiatives. This conclusion represents an important shortcoming in the capacity of private voluntary programs within the financial sector to deliver the incentives necessary for businesses to support decarbonization.

International voluntary initiatives as a broad environmental strategy

The shift toward private voluntary programs in recent decades represents a profound transformation in regulatory regimes that aim to shape the social and environmental performance of corporate actors in the global economy. The 1970s and 1980s saw a strengthening of state-based environmental regulations in many countries. Incidents of poor environmental practices surfaced in a number of industrialized countries in the 1970s, prompting the adoption of more stringent and systematic environmental regulations under what was known as a "command and control" model, which embodied performance standards with significant monitoring and enforcement by the state. By the 1980s, many developing countries started to adopt similar types of environmental regulation. But, with the rise of neoliberal economic policies in the West, and subsequently in many developing countries in the 1980s and 1990s, many states moved away from strict, state-based environmental regulations which were seen as costly and inefficient. Those promoting environmental sustainability via voluntary programs for corporate self-regulation began to gain ground.

The idea of voluntary environmental programs was heavily promoted at the 1992 Rio Earth Summit. Many of the world's leading corporations signaled their commitment at that meeting to "change course" and to "green" their operations voluntarily (Chatterjee and Finger 1994; Schmidheiny 1992). Agenda 21, the main document arising from the Earth Summit, included a chapter on the benefits and potentials of voluntary programs. Although a number of state-based environmental agreements were also negotiated at that time, voluntary programs were viewed as consistent with the broader goals of sustainable development. The performance of voluntary programs broadly has been mixed (as discussed below), but they were again prominently endorsed at the Johannesburg World Summit on Sustainable Development in 2002 (Clapp 2005; Rutherford 2003). Public–private partnerships, arrangements involving private firms in partnership with civil society and public bodies, were promoted as a key means for the furtherance of international environmental goals (Backstrand 2006).

The case for voluntary programs has been made on the grounds that it makes "good business sense" (Holme and Watts 2000; Schmidheiny 1992). Adopting mechanisms that advance resource use efficiency and waste minimization not only cut back firms' costs, but provide an opportunity to promote a greener image and draw new customers, access new markets and attract investors. These benefits are attractive to firms because they can ultimately translate into improvements in their economic bottom line even if they involve initial cost outlays. In this sense, firms are self-motivated to make environmental improvements without being forced into doing so by state-based regulation. Because firms know best where efficiencies can be made within their own business structures, allowing firms to strive for improvements on their own terms is widely seen to be a more efficient way of achieving environmental sustainability.

Throughout the 1990s, governments began to embrace the idea of corporate voluntarism as an approach that was not only more efficient for firms, but also cost-saving for governments, as it meant fewer state resources had to be expended on monitoring and enforcement of government regulations. Some environmental non-governmental organizations (NGOs) that were critical of inefficiencies of state-based environmental regulations also began to support voluntary environmental programs, particularly those involving multiple stakeholders, including civil society. They had hope for the emergence of "civil regulation", which would embody social concern and work directly with those whose behavior they were aiming to change (Vogel 2008; Zadek 2001).

This political opening for private voluntary programs quickly filled with a growing number of international instruments seeking to promote corporate environmental good practice (UNEP 2011). The number of these programs is large and diverse, with numerous initiatives adopted annually, especially in the past decade. Many private programs aim to provide both guidance and standardization to corporate greening efforts at the international level. Some of these programs have emerged from within the corporate world, and are wholly self-initiated and monitored. Others involve non-corporate actors (civil society, environmental NGOs, international organizations, and even states) as participants or overseers, although all voluntary programs work in close cooperation and coordination with the corporate sector. Because of the high degree of corporate participation in such initiatives, including those initiated from "outside" by non-corporate players, many have seen these developments as a kind of "privatization" of global environmental governance, but with a complex relationship to government and other forms of authority (Clapp 1998; Falkner 2003).

The various international voluntary environmental initiatives each address a different need and niche with respect to environmental issues and problems, and as such the landscape has become rather crowded. In order to make some sense of this widening field of voluntary programs at the international level, we categorize them here into four main types that are useful in this particular analysis of the financial sector and its role in climate change politics. Although each is unique and may display characteristics from more than one of these categories, we have placed them where we think they best fit in order to provide clarity.[1]

Principles and codes

Perhaps the broadest category of international private voluntary program is adherence to sets of principles or codes of conduct. These types of measures are not typically monitored or enforced, and firms sign on to demonstrate their commitment to corporate social and environmental responsibility. Many, but not all, voluntary principles and codes that promote sustainable development involve actors from outside industry, such as UN bodies or NGOs, and as such multiple stakeholders are involved. These outside actors often play a key role in the designing and monitoring stages, and in encouraging firms to join them. Having other actors involved in the establishment and oversight of these programs can give them a degree of legitimacy that would be lacking if they came from the industry itself. Principles and codes can be general in nature, applying broadly to all firms, or they can be specific to a particular industry or sector. Codes and principles are popular with firms because they are voluntary in nature, involve little oversight, and are relatively inexpensive to sign up to. By the late 1990s, most large transnational corporations had signed on to one or more of these types of broad initiatives (Utting and Clapp 2008).

Perhaps the most prominent example of a voluntary set of principles at the international level is the UN's Global Compact (GC). The GC is essentially a pact between the United Nations and multinational corporations, which calls on global firms to adhere to a set of 10 principles relating to social, environmental and human rights goals. Launched in 2000, the GC specifically asks firms that sign on and become members of the compact to incorporate the 10 goals into both their mission statements and their general operations (Therien and Pouliot 2006; UN Global Compact 2008a). The specific environmental goals include adoption of the precautionary principle, the promotion of environmental responsibility, and the development and diffusion of environmentally friendly technologies, all issues that have a bearing on climate change. Firms that participate must make a public pledge that they will adhere to these goals. At present there are some 4700 firms that participate in the GC. In its early years, the GC lacked strict oversight to ensure that member firms were in fact adhering to the goals. New integrity measures were added in 2005 which required firms to communicate their progress, and firms that failed to do so were deemed to be "inactive" (UN Global Compact 2008b). Less prominent but also important are the principles of the Coalition for Environmentally Responsible Economies (CERES). The CERES Principles are a 10-point code of environmental conduct for corporations involving multiple stakeholders (Pattberg 2005). Some 70 firms have signed onto these principles, which are somewhat more stringent and involved than the Global Compact, because they focus more on environmental performance and accountability (Brown, de Jong and Lessidrenska 2009).

Other sets of principles focus on specific sectors or issues. The Principles for Responsible Investment (PRI) and the Equator Principles are multi-stakeholder initiatives that set guidelines for the financial and investment industries (e.g. Misbach 2004; Wright and Rwabizambuga 2006). In the chemical sector, Responsible Care (RC) is a global industry initiative that sets out broad environmental, health and safety principles in chemical firms (Lenox and Nash 2003). The Responsible Care standards are set by the International Council for Chemical Associations, an industry group, without much participation from outside stakeholders (Prakash 2000). In the mining sector, the International Council on Mining and Metals (ICMM) established Principles of Sustainable Development in 2003 as an industry-based initiative without participation from a wide range of non-corporate stakeholders (ICMM 2003; see also Sethi and

Emelianova 2006). Other principles, such as the Climate Risk Action Plan, the Climate Principles and the Carbon Principles, are specific to particular environmental problems, and will be analyzed below in relation to the financial sector.

Reporting and disclosure schemes

Reporting on social and environmental performance is another type of private voluntary program that puts particular emphasis on the formalization of various monitoring mechanisms. In the second half of the 1990s, many large firms began to issue regular corporate social responsibility (CSR) and sustainability reports (Lamberton 2005). The idea behind these initiatives is that self-disclosure of information by firms on their environmental and social performance puts pressure on them to ensure that their activities have positive impacts (Gupta 2008). Because these reports are often public documents, shareholders and society more broadly are able to keep tabs on firms' performance. Early CSR and sustainability reports, however, were largely the product of corporate self-monitoring with little oversight or verification of the information. Firms may choose to report information on positive performance with respect to some issues but ignore others where their performance is weak.

With the aim of providing a standardized template for sustainability and disclosure reporting, the Global Reporting Initiative (GRI) was established in 1997. The GRI is a joint project of CERES and the UN Environment Program (Brown et al. 2009). The first set of GRI reporting guidelines was released in 2000, and some 50 firms signed on to adhere to its reporting template. The guidelines were revised in 2002 and again in 2006. Today, over 1000 firms adhere to some degree to the GRI guidelines, although there are different application levels and firms can choose to adhere to an earlier version rather than the most recent. Firms self-declare their level of adherence and this can be externally verified by a third party; however, this verification is not mandatory. In 2002, the GRI became a stand-alone organization and began to revise its standards. The requirements have now increased for new firms signing up to the standards. In addition to the GRI as an overall reporting framework, there are other more specific types of disclosure initiatives at the international level, including those that focus specifically on GHG emissions, such as the Carbon Disclosure Project, discussed below.

Environmental management systems standards

Environmental management systems (EMS) standards set regular management practices within firms. EMS standards are somewhat more involved than broader sets of principles and disclosure frameworks because they often include a certification process. EMS standards, such as the European Union's Eco-Management and Audit Scheme (EMAS) and the International Organization for Standardization's ISO 14000 standards, are typically the product of multiple stakeholders, and certification and auditing of those standards is typically handled by a third party. Firms that sign on to EMS standards are required to standardize their practices in ways that ensure that they keep close tabs on their environmental impact and performance and are also required to commit to improvements.

The ISO 14000 series of standards are perhaps the best-known of the international EMS standards. They were established by the International Organization for Standardization (ISO) and a growing number of firms around the world have become certified (Prakash and Potoski 2006). The ISO 14000 series of standards provide guidance on a number of fronts, and the ISO 14001 standard, on environmental management systems, is the only one to which firms can actually certify their facilities. ISO 14001 requires that firms' management is aware of all relevant environmental regulations and that those regulations are followed in the countries in which they operate. As of 2006 over 154,000 firms operating in 138 countries had certified to the standard (ISO 2007). The ISO environmental management system series of standards added a voluntary GHG standard (ISO 14064) in 2006 and the ISO is also currently working on a new set of standards for corporate energy use known as ISO 50001. These initiatives are discussed more fully below.

Market-based measures

Market-based measures also operate as private voluntary programs at the international level. These include schemes for the certification of particular products to specific environmental and social standards as well as the development of alternative trade and markets that seek to promote social and environmental goals (Utting 2005). The idea behind these initiatives is to create market incentives for firms to improve environmental practices at key points along the supply chain (Cashore 2002). Given the way these initiatives work, many market-based schemes are specific to a particular industry or product. These initiatives may be controlled entirely within a particular industry or have involvement of NGOs or other non-state actors as independent input and oversight providers. As a result, market-driven voluntary corporate initiatives have a wide diversity in terms of the role of independent stakeholders such as NGOs and civil society groups working alongside industry in setting standards, providing monitoring as well as enforcement functions. For this reason, some call these "non-state market driven" (NSMD) initiatives (Cashore 2002).

Alternative markets and sustainable product certification schemes have been established, for example, for organic and fair trade products, sustainable forest products as well as sustainable seafood products (Cashore 2002; Islam 2008; Raynolds 2000, 2004). Supermarket chains are increasingly requiring certification of agricultural produce to "good agricultural practices" through the Global GAP program (Fuchs, Kalfagianni and Arentsen 2009). Because of the range of diversity with respect to independent stakeholders' roles in standards-setting, monitoring and enforcement, some of these schemes are deemed to be stronger than others. New forms of voluntary market-based and product certification schemes have also emerged as a way to address specific environmental issues without being tied to a tangible "sustainable product". In particular, there has been phenomenal growth in market-based approaches to addressing climate change, including voluntary cap-and-trade and project offset schemes. These initiatives will be more closely examined below.

The four main types of private voluntary programs outlined here share similarities in that they are voluntary on the part of firms, involve multiple stakeholders, and have been designed with consultation and participation by the key actors – corporations – whose behavior they aim to change. But these various types of programs also have important differences, employing a range of possible ways firms can "voluntarily" work toward environmental improvements. Although these initiatives are unique each filling a particular need and niche, there are important overlaps and synergies between them. There are links, for example, between the GRI and the GC, and between ISO 14001 and RC. Similarly, alternative markets rely on information available from other types of schemes, such as reporting and management system certification. In this way, many of these initiatives reinforce one another.

Pitfalls of private voluntary programs in practice

Although there has been a great deal of enthusiasm for private voluntary programs since they first emerged (see Esty and Winston 2006; Schmidheiny 1992), practical experience with these initiatives over the past few decades has shown that there are some key aspects on which they fall short. These weaknesses build on the framework established in Chapter 1 by looking at individual "pitfalls" that emerge within the designing, joining, monitoring and complying stages that can compromise the effectiveness of private voluntary programs. The problems include such issues as insufficient implementation and enforcement measures, prioritization of cost-saving measures over others, the tendency to track procedures rather than performance, high rates of complexity, poor participation rates, and the potential to exacerbate inequalities between rich and poor country firms. Together these weaknesses create two serious and related problems for voluntary programs: they highlight the lack of sufficient accountability on the part of firms, and hamper the programs' effectiveness in terms of tangible environmental performance. The presence of these problems raises serious questions about the viability of private voluntary programs as a key anchor for the international promotion of environmental sustainability, and they will be discussed in relation to the schemes now in place in the financial sector.

The overall *poor rate of participation* in private voluntary programs has been flagged as a serious weakness. With all the hype around these programs – and the tendency of some major transnational corporations to highlight their participation in one or more of them – one might assume that most firms adhere to one or more voluntary programs. But, in actual fact, relatively few firms participate. Weak participation represents an important shortcoming in the designing and joining aspects of private voluntary programs. Without broad stakeholder participation, the design of the program may reflect particular interests of those already involved and limit incentives among other firms to join. Globally, there are approximately 70,000 transnational corporations and 700,000 affiliate firms in addition to several million suppliers (UNCTAD 2006). By comparison, there are approximately 154,000 firms certified to ISO 14001, 1000 firms that report according to some form of the GRI guidelines for sustainability reporting, and 800 firms certified by the Forest Stewardship Council (Utting and Clapp 2008). Similarly, the size of alternative markets for fair trade, organic and sustainable products is relatively small. Although these markets may be growing rapidly, they still only make up a tiny proportion of global trade in their product category. Organic food sales make up only 1–2 per cent of global food sales, for example. Because of these poor rates of joining, it is difficult to see voluntary programs, at least at this point in history, as being "transformative" in terms of global environmental governance.

A major critique of private voluntary programs is that they tend to have very *weak implementation and enforcement procedures* and, as such, monitoring and complying fall short. Because they are voluntary in nature, the requirements for joining are often minimal. As such, there is a structural weakness in that requirements remain weak in order to encourage enough participants to take part. Indeed, those programs that have tougher implementation and enforcement procedures, such as the CERES Principles, have far fewer firms signed on than those that are relatively weak, such as the GC. Moreover, most voluntary corporate initiatives are not legally binding, meaning there are few if any serious sanctions for non-compliance. Again, the GC is a good example. It has been virtually impossible for the GC to "kick out" firms that have signed on to the principles but which have had poor environmental performance (Utting and Clapp 2008). Similarly, RC and the ICMM Sustainable Mining Framework, both industry-driven initiatives without multi-stakeholder participation, have been heavily criticized for lacking sanctions against those firms that flout requirements (King and Lenox 2000; Sethi and Emelianova 2006).

A further critique is that many voluntary programs *prioritize cost-saving measures over more costly environmental measures*. Requirements to improve energy efficiency, for example, have direct cost saving benefits for firms and can be advertised as a means by which firms are taking action to not only conserve energy, but also cut back on GHG emissions. But activities in other areas may not have direct cost saving benefits and these activities generally tend to be less prominent in voluntary initiatives. For this reason, complying with such standards often generates few real environmental benefits. To take a specific example, the GRI includes explicit requirements for firms to report on activities to mitigate climate change and conserve energy, but little is required in terms of reporting on activities undertaken by firms to slow the spread of genetically modified organisms (GMOs) to jurisdictions where they are not approved (Clapp 2008). The tendency to focus on cost-saving environmental measures may make sense as an initial focus as a way to get firms to join. However, it does raise questions about the value added of these programs if they only encompass measures firms would undertake in any case, while other issues remain "off limits" (Utting 2008).

Private voluntary programs have also been critiqued because they typically *do not measure concrete outcomes in terms of environmental performance*, but rather focus on the standardization of goals, management practices or reporting procedures. A number of principles and codes call on firms to commit to "continual improvement" but do not set minimum performance standards (Sethi and Emelianova 2006). In this circumstance, initiatives that support robust monitoring and compliance may fail to succeed in achieving any improvements in a firm's environmental performance. Certification to the ISO 14001 standard, for example, only verifies whether aspects related to environmental management systems are in place, and not whether these practices have had any concrete outcome in terms of improved environmental performance (Krut and Gleckman 1998). In other words, firms that sign on to ISO 14001 pledge to ensure that their management is aware of how they perform environmentally, but are not required to meet any specific standard of performance (Clapp 1998).

Despite the lack of stringency, there is still a great deal of *complexity associated with voluntary programs*. The sheer volume of voluntary programs operating simultaneously, with some addressing the same niche, creates significant amount of confusion. This complexity makes monitoring and control of actual compliance difficult because external auditors, or the firms themselves, are unable to determine if, compared to firms in other programs, they are actually improving their environmental performance. With so many initiatives in play, in many cases firms can pick and choose which ones they join. Indeed, as Chapter 1 describes, firms find it easier to join already established programs without participating in the original negotiations. These firms can then "free-ride" on the reputational gains generated by the firms that took the initial steps to design, join, monitor and comply with the program. It is difficult for those tracking the environmental practices of firms, let alone the general public, to keep up with the various programs, their requirements, and which firms participate. Even within particular programs, complexity is common because firms often have a variety of actions they can take. For example, each time the GRI introduces a new and updated set of reporting guidelines, it allows firms to adhere to previous versions and still be certified. Further, there are different degrees of strictness with which firms can follow these guidelines. When high degrees of complexity are present, the certification at the end of the day loses its meaning. There is also the risk, in particular with disclosure-type mechanisms, of "drowning in disclosure" when the information is voluminous and in a format that is difficult to decipher (Gupta 2008).

There may also be *North–South inequities* associated with the growth in popularity of voluntary programs. There are fears, for example, that a number of such programs undertaken by industrialized country firms may work against firms from developing countries that do not join, as they may be shut out of the supply chain. Although the overall costs of joining these programs may be relatively low in industrialized country firms, the cost can be quite high for small and medium-sized enterprises and producers operating in developing countries in particular (Fuchs et al. 2009; Nadvi and Wältring 2002). This problem is especially acute for those voluntary programs that require third-party auditing and certification. It is typically the case that auditors from developing countries do not have the same recognition abroad, and thus developing country firms may be in effect required to pay for third-party auditing from industrialized countries at much higher cost to them than would be the case with local auditors (Clapp 1998). As a consequence, embedded North–South inequities can create disincentives in the joining stage for firms from developing countries.

Some of these weaknesses apply more to certain types of voluntary programs than to others, but because the various programs all relate to one another in complex ways, they have broad relevance. Further, these various weaknesses tend to reinforce one another in the designing, joining, monitoring and complying stages involved in developing a successful private voluntary program. For example, self-reporting initiatives are supposed to work by leveraging a firm's corporate reputation to improve environmental performance by providing transparency on their ecological footprint to various stakeholders who can then direct attention to firms with increasingly dangerous environmental impacts. However, if a reporting initiative only discloses data on the basis of various monitoring activities, and provides no enforcement mechanism for reducing their ecological footprint, the firm has no incentive to take any additional action beyond simply tracking its own pollution and monitoring its corporate reputation.

When considered together, these various problems associated with private voluntary programs at the international level present serious challenges in their ability to be reliable tools of global environmental governance. First, they highlight serious gaps in *accountability* in the various measures. Second, they put a significant dampener on the environmental *effectiveness* of these measures. As a result, these problems add up to a lack of incentive for firms to go much beyond "business as usual". Indeed, the design of all the voluntary programs outlined above included significant representation by corporate players at all stages of their development and operation, and as such they are widely seen to be measures that only go as far as firms are willing to go within their own economic interests; this is why many critics have labeled them as "greenwash" (Bruno and Karliner 2002).

Voluntary climate change programs and the financial sector

In recent years, the financial sector has engaged with a variety of stakeholders in several related processes to design and participate in a range of voluntary climate change programs. These programs deserve more attention because they have potentially broad, cross-sectoral implications.

Individually, these programs are designed to standardize financial practices and methodologies within the financial sector, or within firms in the broader global economy, around the creation of a cost structure on GHG emissions. Mostly, they are governed using a "hierarchical" monitoring system where an external actor works with firms to design the program, encourage participation and monitor compliance. But, in aggregate, these programs represent an effort to generate a coherent market signal on the financial performance of firms under pressures associated with a GHG-constrained economy. In other words, the end goal of many of these initiatives is to create a "horizontal" monitoring system where firms compete in the implementation of standards that reduce their exposure to climate change risks. For example, voluntary emissions trading markets are governed by standards generated by an external actor that puts a price on carbon at the firm level. This price creates an incentive to reduce emissions, and provides a market signal on the firm's financial performance in a GHG-constrained economy. This signal informs investment decisions and the levels of capital costs and premiums set by the financial sector, which then imposes another set of costs on the firms' GHG emissions. Because these programs have the potential to impose costs on GHG emissions across the entire private sector, an analysis of their capacity to be effective in generating these cost structures is necessary.

Principles and codes: The Climate Risk Action Plan and the Climate and Carbon Principles

Although there are a variety of established environmental principles and codes of conduct in the financial sector, only in the last few years have specific climate-related voluntary programs emerged. The idea behind climate-related principles in the financial sector is that banks and other investors can impose a cost structure on GHG emissions by leveraging their control over access to financial resources. The Climate Risk Action Plan ("Action Plan"), the Climate Principles and the Carbon Principles are three examples of programs that financial institutions can use to evaluate and reduce the impact their investments have on climate change. Four important weaknesses, also pointed to in the existing literature, emerge based on analysis of these principles, including complexity and redundancies between initiatives, failure to prioritize environmental performance, weak rates of participation and a lack of enforcement.

The Investor Network on Climate Risk (INCR), a group of institutional investors managing over US$7 trillion in assets, and the Coalition for Environmentally Responsible Economies (CERES), partnered to design the Action Plan in 2008 (INCR 2008a). CERES hosts the INCR staff and monitors compliance among investors to the plan. Signatories to the plan include pension funds, treasurers and asset managers from various US states and European countries. The plan outlines a series of steps institutional investors can take to encourage both investors and the firms they invest in to begin costing the risks and potential opportunities afforded by a low carbon economy. There are nine standards in the Action Plan which include, for example, a requirement to screen their investment managers on how they are assessing climate change risk, an encouragement to invest in companies developing clean technology, and measures to improve the energy performance of real estate and investment portfolios. The standards further ask members to pressure the firms they invest in to develop comprehensive responses to climate change through full disclosure of emissions and expand the use of climate change risk management. They also leverage the network to support policy action (INCR 2008b).

The Climate Principles are the result of a partnership between the Climate Group, an environmental coalition between cities, states, governments and corporations, and several large financial firms including HSBC and Munich Re. Similar to CERES's role in the INCR, the Climate Group monitors implementation and compliance to the principles. These principles follow a similar framework to the Action Plan by attempting to set a performance standard for financial institutions. More specifically, they attempt to minimize the financial sector's operational carbon footprint; develop products that their clients can use to manage climate change risks; expand dialog with clients to develop financial opportunities from a low carbon economy; publicly support energy and climate change policy; and report on their progress towards the commitment (Climate Group 2008).

Similar to the Climate Principles, a multi-stakeholder process was involved in developing the Carbon Principles. The Carbon Principles were developed over a nine-month period in 2007 through consultations between Citi, JP Morgan Chase, Morgan Stanley, several large US power companies and environmental NGOs. The Carbon Principles operate somewhat differently than either of the previous two sets of principles. Instead of targeting firms to adhere to a set of performance standards, the Carbon Principles create a common approach, called the Enhanced Environment Diligence Framework (EEDP), which evaluates the risk of GHG costs when investing in GHG-intensive activity (Carbon Principles 2008). Although NGOs provide advisory support to the signatories, there is no formal "hierarchical" monitoring mechanism, as is the case with the Action Plan or Climate Principles.

The menu of choice between all three sets of principles reveals the first weakness: that there is a great deal of complexity and redundancy between these programs. Firms can pick and choose which set of standards they wish sign onto based on the least cost to their operations. For NGOs or other concerned stakeholders monitoring a firm's climate change efforts, it is difficult to monitor compliance or compare the performance of two firms signed into different principles and examine the degree of compliance. Whether a signatory to the Action Plan, which reduces the energy usage in their real estate investments, or a signatory to the Climate Principles, which offers an investment portfolio that channels money to renewable energy, is doing more to address climate change, is really anyone's guess.

Both the INCR's Action Plan and the Climate Principles address the common pitfall among principles and codes that they often fail to target environmental performance. In particular, the Action Plan asks signatories for a 20 per cent reduction in the energy used in their real estate portfolios (INCR 2008b). The Carbon Principles, however, are an example of a set of principles that fail to target environmental performance. Instead, the principles provide a due diligence framework to evaluate the financial risks of investments in carbon-intensive sectors, rather than asking for an actual shift in investment strategy to cleaner energy (Carbon Principles 2008).

This due diligence procedure should in theory curb investment in carbon-intensive economic activity that represents a significant financial risk. However, according to BankTrack, there is little evidence that signatory banks to the principles have reduced financing for carbon-intensive industry such as coal power (BankTrack 2008). Located in the Netherlands, BankTrack is a network of various environmental NGOs, including the World Resources Institute, Friends of the Earth and the Rainforest Action Network, which monitors the private financial sector and its environmental impact. Although BankTrack is not formally involved with the Carbon or Climate Principles, it has gained considerable experience monitoring the Equator Principles, another set of environmental principles in the banking sector.

Although participation rates for the Action Plan are impressive, the Carbon and Climate Principles suffer from weak rates of participation. There are 50 international signatories to the Action Plan, representing portfolios worth US$1.75 trillion. Some notably large institutional investors are members of the group, including AIG Investments, BlackRock Financial, CaLPERs and Deutsche Asset Management. The Climate and Carbon Principles only have five (Credit Agricole, HSBC, Munich Re, Standard Chartered and Swiss Re) and six (Bank of America, Citi, Credit Suisse, JP Morgan Chase, Morgan Stanley and Wells Fargo) signatories, respectively.

Because only a few large and publicly visible institutions have joined these programs, they reveal another weakness: that the motivation for participation is likely cost savings, or more specifically, the mitigation of reputational costs (Wright and Rwabizambuga 2006). To shore up their corporate reputation, for example, firms that sign up to either set of principles enjoy endorsements from important environmental NGOs including Environmental Defense, the National Resources Development Council and CERES (Carbon Principles 2008; Climate Group 2008). Similarly, firms that sign on to the Action Plan gain an endorsement from CERES (INCR 2008a). By employing a due diligence mechanism to reduce the exposure to potential GHG costs, the Carbon Principles can certainly be criticized for prioritizing cost savings over the environment. According to BankTrack, the Carbon Principles do little more than buffer banks from potential regulatory risk by incorporating GHG emission prices into the future operating costs of large emitters (BankTrack 2008).

Finally, a weakness of all three programs is the lack of an enforcement mechanism that controls and secures compliance, which significantly hurts their effectiveness and contributes to an accountability deficit for signatories. Although both the Action Plan and the Climate Principles set out financial sector performance standards, there is no way of knowing whether these firms are actually meeting them. The Climate Principles do provide a disclosure mechanism where signatories are obliged to monitor and list their progress. However, this disclosure is not enforced, and when firms fail to report their progress, they are merely "encouraged to disclose a timeframe for achieving full implementation" (Climate Group 2008). Similarly, the Carbon Principles' due diligence framework does not have to be implemented uniformly, and banks can decide when to use the procedure and, ultimately, determine whether the information it produces is relevant. As a result, the Carbon Principles fail to provide any accountability mechanism. More significantly, none of the three sets of principles provides any indication on the emissions produced by the signatory's investments. The Climate Principles take an important step to investigate their signatories' emissions, but only through secondary sources that remain unverified, such as the Climate Disclosure Project (CDP), corporate reports and websites (Climate Group 2008).

Based on analysis of INCR's Action Plan, as well as the Carbon and Climate Principles, it is difficult to argue that the signatories to these programs will shift their business-as-usual behavior, or that of their clients. Patchwork levels of participation, complexity between the principles, prioritizing corporate reputation over environmental performance and defective levels of enforcement contribute to problems in both effectiveness and accountability. In reference to the framework established in Chapter 1, these patchwork levels of participation reveal the challenges involved in convincing firms to join and, more importantly, take part in the design of the initiative. In addition, complexity, prioritizing corporate reputation over environment, and weak enforcement mechanisms limit the monitoring and compliance necessary to hold firms accountable in reducing their exposure to climate change risks or reducing emissions. In this respect, these programs fail to produce any outcomes that can be considered climate change mitigation in the concrete sense, and the theory that the financial sector can impose costs on access to capital as a motivation for change in other industries is put in doubt. BankTrack provides an assessment for all these principles and argues that "even if they were to be adopted tomorrow by every financial institution in the world, they would only result in, at best, a moderate departure from existing business practices" (BankTrack 2008).

Reporting and disclosure: The Climate Disclosure Project

The Climate Disclosure Project (CDP) was initiated by a group of 22 UK institutional investors in 2000. Their goal was to collect information on the climate change impact of the world's largest corporations. Each firm signed a survey which was sent to CEOs from the world's Fortune 500 companies asking specific questions on each firm's GHG emissions. Paul Dickinson and Tessa Tennant – two policy entrepreneurs with experience working in the UK government and financial sector on issues of corporate sustainability – spearheaded the initiative. After establishing the CDP Secretariat to design and monitor the survey, which included various representatives from the investor community in addition to the two founders, the CDP grew rapidly in the years after the initial survey, expanding the number of international signatories and respondents to its survey. By asking former US President Bill Clinton and other high-profile figures to present the CDP results after each annual survey, the CDP has gained a great deal of legitimacy as the world's premier benchmark on GHG emissions (Spencer and Cui 2007).

The goal for the CDP was to produce a standard reporting metric that provides information to guide investment based on a firm's exposure to climate change risks as well as opportunities in a low carbon economy. Similar to climate change principles and codes, the CDP represents an organized effort by the financial sector to leverage access to private investment capital as an incentive to disclose emissions, and subsequently, reduce emissions (Kolk, Levy and Pinkse 2008: 721). While the CDP enjoys a high level of participation, there are several weaknesses that limit its capacity to leverage investors in promoting disclosure and GHG mitigation by the firms they invest in. The CDP has no enforcement mechanism for disclosure, fails to target environmental performance, suffers from low levels of participation from firms in developing countries, and discloses information that is inconsistent and too complex to represent a standard measurement on emissions.

Each year, the CDP Secretariat, which is registered as a charity in both New York and London, sends out a "request for information" to the chair of the board of the world's largest companies across a huge variety of industries. This request contains a wide array of questions on a firm's ability to mitigate GHG, but most importantly, targets full disclosure of their emissions (CDP6 2008: I; CDP6 2008: 2). Firms then voluntarily decide on how much information to disclose, based on three different measures for GHG emissions. Scope 1 emissions count the direct combustion or release of GHG from the "end of the pipe". Scope 2 emissions cover the GHG emitted further upstream from the purchase of energy. Scope 3 emissions are much more difficult to count because they cover abstract sources such as business travel, external logistics, transport, supply chain and product use (CDP 6: 117). By reporting this data on emissions, investors have three comparable benchmarks on emissions levels they can incorporate into due diligence procedures that evaluate the capital costs on their loans, or risks in their portfolios.

Increasing and wide participation in the CDP from the world's largest corporations represents the survey's most important strength. Focusing on response rates in the world's largest 500 firms, or FT500, the CDP has enjoyed an increasing response rate, rising to 409, or 82 per cent of these firms. Although there is no obligation for investors besides acting as a signatory, the CDP has also gradually increased its signatories to 475 investors holding over US$55 trillion in assets (CDP 2009: 8; Kolk et al. 2008: 740). From this perspective, the CDP has demonstrated success in the joining stage in developing a private voluntary program.

The most significant weakness of the CDP is its failure to enforce the disclosure of important emissions data. Besides latent investor pressure, there is no actual mechanism forcing firms to monitor and disclose their emissions. For the CDP to be effective, investors need quantifiable data that is standardized for all firms involved in the disclosure project. Unfortunately, there is no way of accounting or confirming that the response from firms involved in the survey actually reflects their emission levels (Kolk et al. 2008: 743). Of the respondents to the 2009 survey, only 49 per cent disclosed emissions that were verified by a third-party auditor (CDP 2009: 10). Most of these responses are from firms already participating in the EU Emissions Trading System (ETS) and other regulated emissions trading systems, where disclosure verified by a third-party auditor is mandatory (Kolk et al. 2008: 738). For a firm operating inside a regulated emissions trading market, withholding data on emissions is subject to a fine, whereas the CDP provides little incentive to develop the necessary measuring or accounting framework that can track emission levels (CDP5 2007).

This lack of enforcement leads to a complex patchwork of emissions data because firms can pick and choose how to respond, and, consequently, complying becomes fragile and uncertain. Such complexity constitutes a second weakness in the CDP. The CDP has tried to adjust its strategy by working with investors to improve the design of the survey. For example, more recent surveys have focused on more rigorous quantitative questions, such as cost savings from emissions trading programs. The response to this effort produced a financial figure from a meager 27 firms. This complexity is, however, compounded by the constantly changing design of the CDP survey. Each year new questions or sub-questions are added, which limits the comparability of results between years and produces an inconsistent metric for investors or NGOs evaluating emissions mitigation. Moreover, these questions often prove too onerous for respondents, who must constantly update their store of data on climate change strategy. From the perspective of firms also, then, complying is difficult to manage properly (Kolk et al. 2008: 735).

Although overall participation rates seem high, North–South inequities are apparent as developing country firms have low participation rates due to the high costs of adhering to the disclosure standards. For firms in developing countries, establishing effective GHG accounting is an expensive operation that often requires the assistance of Northern auditors. This disadvantage for firms in developing countries is reflected in the CDP's responses, where very few firms from developing countries participate or provide adequate disclosure. For instance, in 2009, 70 per cent of the responses came from firms in five industrialized countries: France, Germany, Japan, the US and the UK (CDP 2009: 25). This trend is consistent with the 2008 CDP where only 14 per cent of Indian and 15 per cent of Chinese firms responded to the request for information (CDP6 2008: iii).

With weak enforcement, patchwork and complex reports of emissions data, varying levels of participation and North–South inequities, there is no way for the CDP to hold firms accountable for the information they provide. These weaknesses break down individual obstacles within the "designing, joining, monitoring and compliance" aspects to effective voluntary private programs. The CDP's effort to consistently improve the design of its survey increases complexity, which complicates efforts to monitor compliance among firms disclosing their emissions. In addition, varying levels of participation demonstrate the challenge involved in convincing firms to join the survey. Outside of these regulatory trading schemes that legislate third-party auditing, there is little evidence firms will voluntarily take up the challenge to consistently monitor and disclose their emissions. More fatal, however, is recent evidence that investors often ignore information from GHG disclosure to guide their investment decisions (Gray, Murray, Sinclair and Power 2006; Kiernan 2008). In this respect, not only the CDP's disclosure mechanism, but also its strategy of using access to investment capital as a lever to reduce emissions, remain unproven.

Environmental management systems: ISO 14064 and 50001

Both ISO 14064 and 50001 represent efforts to change common business practices by adopting an accounting system that measures a firm's emissions and its energy usage. In this sense, they are promoting practices of the financial sector, and the accounting of carbon and energy uses in turn has important implications for investors. Although similar in design to disclosure frameworks, EMS standards attempt to go one step further not only by implementing a standardized accounting mechanism that measures GHG emissions, but also by internalizing the cost of emissions by setting performance targets for reductions at the firm level. Despite the difference, ISO EMS standards adopt a similar approach to other initiatives by using a "hierarchical" monitoring approach with the long-term goal of generating a "horizontal" structure where firms compete with each other in measuring and reducing their GHG emissions. ISO 14064 and 50001 have the potential for wide coverage within the private sector, but there are several important pitfalls that reflect problems common to voluntary EMS more generally. Most notably, ISO 14064 and 50001 suffer from several of the weaknesses pointed to in the literature on private arrangements, such as weak enforcement mechanisms, North–South inequities, prioritizing cost savings over other environmental concerns, and failure to set adequate targets for environmental performance.

ISO 14064 was approved for international use in March 2006, but discussions on its design began in 2002 as an attempt to standardize how firms account for and measure their emissions. The standard is divided into three areas: parameters on setting up a GHG inventory; requirements for identifying good sources of GHG offsets; and a framework for verifying the accuracy of emissions in the inventory. The verification procedure is important from an investor perspective because it takes emissions recorded in the inventory report and evaluates whether the data is consistent, sound and accurate. This part also specifies which GHG auditors are credible in counting emissions and preparing validation and verification reports (Boehmer and Cherp 2003).

ISO 50001 was designed to provide a system where firms can assess, manage and measure energy usage. Firms can use this system to measure their use of energy at each step of the production process and integrate energy efficiency into decision-making by management. More specifically, ISO 50001 will inform firms how to calculate a baseline for their energy use, manage energy use and costs, reduce emissions, and produce a common reporting framework that documents savings (Alliance to Save Energy 2008). To apply these measures, the firm uses a "Plan-Do-Check-Act" framework to manage energy resources. Firms have an incentive to adopt ISO 50001 because it harmonizes national energy management systems into one standard, and levels the playing field for firms competing to improve their energy efficiency (ISO 2011).

Evaluating these programs reveals an important strength, but many more weaknesses. As a procedure to standardize accounting of how firms measure emissions and energy usage, ISO 14064 and 50001 have the potential for wide coverage and membership. Over 50 countries were involved in the development and design of ISO 14064, and ISO 50001 could potentially impact 60 per cent of the energy used in the world (Toon 2008; Wintergreen 2007). As the world's premier standardization organization, ISO enjoys a great deal of legitimacy, and the adoption of an emissions and energy usage standard leverages this legitimacy. However, outside its potentially wide coverage, evaluating ISO 14064 and 50001 reveals that they fail to provide any enforcement, do not recognize the technical and financial limitations of firms in developing countries, and do not incorporate environmental performance targets. As a consequence, any effort by the ISO to improve monitoring or compliance is likely to yield few, if any, environmental benefits.

Potoski and Prakash have developed a framework for evaluating the level of enforcement for EMS programs. They argue that "strong" enforcement for EMS programs must include "third party audits, public disclosure and sanctions" (Potoski and Prakash 2006: 20). The level of enforcement for both ISO 14064 and 50001 would be considered "weak" according to Potoski and Prakash because this information is not publicly available and there is no sanction for non-compliance. In contrast, the EU's EMAS requires firms to promptly disclose information to the public, identifies those firms not providing the minimum level of disclosure, and even establishes performance targets that firms must achieve to remain EMAS accredited (Gray and Bebbington 2001: 107).

Fulfilling the requirements for EMS standards such as ISO 14064 and 50001 may prove to be too complex and costly, specifically for firms in developing countries, which dampens their coverage. GHG accounting is incredibly complicated to design effectively, especially if a firm is simultaneously purchasing GHG offsets, making their own reductions, and selling these reductions in an international market. For example, effectively measuring "scope 3" type emissions, such as plane travel, supply and product chain emissions, is often beyond the capacity of most firms. Because verifying levels of "scope 3" emissions is so complex that the EU has banned including these emissions in their emissions trading market (EC 2004). Considering one of the major reasons firms adopt ISO mechanisms is its international coverage, the inability of many firms in the world to actually participate in such a mechanism is an important limitation.

A third weakness of ISO 14064 and 50001 is that they prioritize cost savings over other environmental concerns. Both mechanisms intend to standardize how firms measure GHG emissions and the amount of energy they use. Firms can then account for and improve their operation's efficiency and save money. By focusing only on energy and GHG emissions, attention is drawn away from other activities such as tracking water pollution and toxic substances. For environmental accountants, a rigorous EMS would not only audit emissions and energy use, but also monitor and account for all pollutants as a measure of a firm's value. By employing a more comprehensive EMS (i.e. more than simply an audit of emissions and energy), firms can make full use of EMS by "learning how to manage environmental performance" (Gray and Bebbington 2001: 99).

The overarching weakness of ISO 14064 and 50001 is that they fail to target environmental performance or GHG reduction, and only focus on standardizing the accounting of emissions. This design approach, while conducive to more firms joining the survey, has the potential to create a perverse outcome as the ISO monitors and improves compliance to effective measurement that generates few environmental benefits. For this reason, environmental accountants routinely criticize EMS as examples of auditing as a "substitute for doing" (Power 1991). Developing a program that only measures emissions and energy usage should not be considered an environmental management system, but rather a disclosure mechanism similar to the CDP. However, even in this role, ISO 14064 and 50001 fail to provide enough information to generate a market signal that can be used by the financial sector to impose costs (e.g. higher interest rates and premiums) in addition to those required to internalize the cost of emissions at the firm level.

Market-based measures: Voluntary emissions trading markets

Compared to other voluntary climate change programs, voluntary carbon markets represent the most sophisticated and organized use of financial practices and knowledge by invoking a market structure as a mechanism to price, and then competitively reduce, GHG emissions. Participating in voluntary carbon markets represents an actual governance mechanism where observable GHG mitigation is not only possible, but investors and financial firms are provided with a market signal to determine which firms are likely to be winners and losers in a GHG-constrained economy. Although described as the "wild west" compared to its regulatory counterparts in the public sector, such as the European Union's Emissions Trading Scheme (EU ETS), or the UN Clean Development Mechanism (CDM), voluntary carbon markets provide a much "thicker" governance effort than GHG disclosure, principles or management systems. Voluntary markets offer firms a "shadow" market where they can practice or develop the management expertise required to participate in regulated emissions trading markets. Despite only constituting a small fragment (2.2 per cent) of total global emissions trading, voluntary markets are increasing at almost double the rate of regulated markets (Hamilton, Bayon, Turner and Higgins 2007: 20).

Until recently there were two different types of voluntary markets in operation, one of them now defunct. The Chicago Climate Exchange

(CCX) was a structured and monitored voluntary cap-and-trade market that enforces a mandatory cap on emissions. Although the CCX was closed in 2010, it represents an important example of a voluntary emissions trading market that can be used to evaluate the effectiveness of these markets. The CCX was designed and developed in 2003 by Richard Sandor, an economist and CEO of Environmental Financial Products, who is famous for his role in the development of futures markets during the 1980s when he was chief economist to the Chicago Board of Trade. For emitters that signed up to the CCX, there was a "voluntary, but legally binding commitment" to reduce GHG emissions by 6 per cent before 2010 from a baseline set by their average level of emissions between 1998 and 2001 (CCX 2009). Failure to meet this requirement results in a revocation of membership of the CCX, and so complying is a key element. Another voluntary market outside the CCX, the "over-the-counter" (OTC) market, operates on a deal-by-deal basis where emitters purchase "GHG offsets" for their emissions. Over 65 per cent of trades in the voluntary market occur through the OTC market (Hamilton, Sjardin, Marcello and Xu 2008). There is little doubt that the organization of a voluntary market for emissions trading is a remarkable accomplishment for voluntary programs. But, under closer scrutiny, these voluntary markets suffer from a series of weaknesses that limit their capacity to mitigate GHG. These weaknesses include weak enforcement mechanisms, patchwork participation, failure to set an environmental performance target, North–South inequity, and complexity between initiatives.

The first three weaknesses are related and in aggregate prevent the market from reducing emissions and improving the environmental performance of firms that participate. For these markets to improve environmental performance, they need to be designed in ways that ensure that each trade of GHG is both an additional and permanent reduction of GHG, and that the price is high enough to punish the firm for subsequent emissions. "Additionality" means that tradable GHG represents a reduction in emissions that would not have otherwise happened. "Permanence" means that once a reduction in emissions occurs, the reduction is permanent (Kollmus, Zink and Polycarp 2008: 15). Without a high price, which most economists argue is between US$30 and US$85 per ton, firms have no incentive to invest in cleaner technologies, because the cost of emitting fails to threaten their revenue streams (Nordhaus 2007). If non-additional, non-permanent, or cheap GHG credits enter a market, firms that purchase these credits as offsets will continue to emit, while no emission reductions have actually been achieved.

Voluntary carbon markets are increasingly prone to the trade of GHG credits that ultimately fail to improve environmental performance, for two reasons. First, unlike regulated markets where participation is mandatory and a cap is used to introduce scarcity, voluntary markets cannot create scarcity because there is no cap in OTC markets, and only those firms that can fulfill the CCX cap tend to participate. As a consequence of this patchwork participation and no enforceable cap, these markets fail to support a price commensurable with a real reduction in emissions (Chan 2009). For example, the volume-weighted price average on the OTC market for 2006 was US$6.1, whereas the price in the CCX was even lower, at US$3.15 (Hamilton et al. 2008: 8). These prices fall well short of providing any disincentive for a firm to adopt cleaner or more efficient technology, and provide incentives for firms to participate in these markets as a mere financial exercise. In fact, large financial institutions, such as hedge funds, often see voluntary markets as a cheap way to speculate on credit prices or hedge their portfolio instead of a tool to reduce emissions (*Carbon Finance* 2009).

Second, without a mandatory cap that forces firms to reduce their emissions through investing in cleaner technology, firms can purchase project offsets that often fail to represent a real reduction in GHG emissions. Project-offset GHG credits are created by measuring a reduction from a hypothetical "baseline" for future emissions (Capoor and Ambrosi 2007: 8). Other emitters, usually already involved in a regulated scheme, can then purchase the credit to offset emissions and satisfy their obligations. In both the OTC and CCX market, project offsets constitute the overwhelming majority of trades (Hamilton et al. 2008: 7). These offsets often fail to represent a real reduction in GHG emissions because emitters can manipulate the baseline by increasing emissions before the measurement, which inflates the baseline, making any subsequent reduction non-additional. Also, many offsets in unregulated markets originate from biomass sequestration projects, which the EU Commission has banned out of concerns that additionality and permanence are compromised (EC 2004; Bumpus and Liverman 2008: 136; Chan 2009). Even in heavily regulated markets, where a reduction in emissions is mandatory, it is difficult to ensure that GHG offsets improve environmental performance. In fact, the US Government Accountability Office argued that, in the CDM market, "it is not possible to ensure that every credit represents a real, measurable, and long-term reduction in emissions" (US Government Accountability Office 2008).

A high number of offsets generated for unregulated markets come from developing countries.[2] Ideally, Northern firms can invest in emission reduction projects in Southern countries that offset their emissions and also benefit the energy infrastructure of local communities. Unfortunately, this design often does more to increase inequity between firms in industrialized and developing countries and disadvantages mitigation efforts of local populations. Because these projects require such a high level of auditing and oversight, most communities in developing countries cannot afford to participate in the unregulated markets. The firms in those countries that can afford to participate are large and often foreign-owned. In many circumstances, these firms are indeed the targets of local protests instead of a source of sustainable development. There is also increasing evidence that project offsets actually contribute to fossil fuel dependence in developing countries. Under even the most regulated project offset markets, there are perverse examples where offsets are generated through reducing carbon dioxide (CO_2) emissions by pumping it into oil deposits to ease access for drilling. Needless to say, using CO_2 offsets to generate both finance and access for oil extraction was not the intended goal for these markets. As a consequence, these firms gain from their participation in the market, whereas communities initiating climate change mitigation strategies through small-scale efforts go unrewarded (Lohmann 2006).

To address North–South inequities, and the failure of the market to target environmental performance, a small industry of third-party auditors has emerged, which act as brokers and intermediaries, guaranteeing that credits are additional and permanent, and also benefit local populations. Of these auditors, the Voluntary Carbon Standard, the CDM protocol developed for the UN's Clean Development Mechanism, and the Gold Standard, are among the most popular (Hamilton et al. 2008). These auditors reveal the emergence of a "hierarchical" monitoring system, where international agencies and even NGOs implement standards that try to improve the design of the market by certifying the quality of the credits traded by market participants. To elaborate on the roles of these auditors, the Gold Standard will be used as an example. The Gold Standard Foundation is a non-profit organization that has developed a "best practice methodology" for producing high-quality GHG credits usable in both unregulated and regulated markets (Gold Standard 2008). Launched in 2003 as a worldwide standard for GHG credits, the Gold Standard was initially conceived by a group of NGOs at the seventh Conference of the Parties (COP) in Marrakech as a complementary certification process for project developers participating in the CDM. The Gold Standard is governed by a foundation board, which is composed of representatives from the over 60 NGOs that support the standard.

When the Gold Standard is used to set up a project, consultation meetings are held with local populations to ensure that concerns about the project's environmental and social impact are heard. As a result of this rigor, over 60 NGOs endorse the Gold Standard's methods for producing a voluntarily created project offset (Gold Standard 2008). Indeed, offsets purchased through the Gold Standard sustain a volume-weighted price of US$17.1, which is closer to the price most economists argue is necessary for behavioral change (Hamilton et al. 2008: 49). Although these third-party auditors and verifiers provide a needed level of quality assurance to the voluntary market, there is no way of confirming that the GHG credits they produce result in improved environmental performance.

Complexity also hampers these initiatives, as firms can pick and choose between a variety of different offsets markets as well as auditors to buy credits from. If a firm finds reducing their emissions through the adoption of technology is cheap, it can join a voluntary cap-and-trade scheme and benefit from the cap-and-trade market. On the other hand, if a firm finds adopting technology to reduce emissions too expensive, it can enter the OTC market and find a cheap source of offsets on their own. For example, a World Wildlife Fund (WWF) report describes over 10 different auditors. Each auditor has its own standard for environmental integrity, which allows firms to choose the standard that is most cost-effective for them (Kollmus et al. 2008). For this reason, both the EU ETS and the newly emerging US carbon market have developed their own auditing systems to evaluate project offsets (EC 2004; US House of Representatives 2009). In this respect, efforts on the part of external auditors to improve the design of the market can lead to problems in terms of monitoring emissions reductions because firms can purchase offsets that are not additional or permanent.

Voluntary carbon markets represent the most sophisticated voluntary climate change program to mitigate climate change, and are not only built on financial sector practices but have themselves become arenas for financial speculation. However, they are prone to critical weaknesses, including patchwork participation, weak enforcement, a failure to target environmental performance, North–South inequities, and complexity between initiatives. Many of these weaknesses relate to design problems inherent in a "voluntary" market. Ineffective oversight and parameters on the credits firms can purchase creates "spill-over" in terms of monitoring whether firms are complying with the objective of the market, which is to reduce emissions by purchasing offsets that "cover" their GHG footprint. Without a transparent administrative structure, investors and financial firms have no incentive to look to these markets to inform investment decisions on which firms will prove competitive in a low carbon economy because there is no way to tell when a firm is actually reducing emissions while maintaining a profit. As a consequence, the capacity of these markets as a voluntary climate change program designed to create financial incentives to reduce GHG emissions remains highly in doubt.

Conclusion

Private voluntary programs represent an impressive development in private self-regulation in terms of their scope and the diversity of mechanisms they employ. They have become enormously popular with firms and industries over the past decade as a way to show their commitment to the promotion of sustainable development. In order to make sense of the rapidly growing number of programs at the international level, we have developed and employed a loose typology including different voluntary elements in building programs. These include principles and codes of conduct, disclosure and reporting mechanisms, environmental management systems, and market-based measures. Although distinctive, each of these approaches attempted to create rules or norms whereby businesses consider their contribution and exposure to climate change impacts. While many have faith in these various types of programs to improve the environmental performance of firms, the literature analyzing their impact in practice has uncovered a number of weaknesses which impede their ability to contribute to a tangible improvement in environmental performance and which also reveal important gaps in accountability.

The application of this typology to voluntary climate change programs in the financial sector addresses an important gap in existing literature on the effectiveness of these programs, and expands analysis on the emerging link between the governance of financial markets and environmental issues. By using this typology to evaluate voluntary climate change programs that embrace the financial sector and its practices, this chapter also builds on the framework established in Chapter 1 by identifying individual weaknesses in the designing, joining, monitoring and complying aspects of effective voluntary private programs. An analysis of climate change principles, disclosure mechanisms, environmental management systems and voluntary emissions trading markets reveals six important weaknesses. These weaknesses include insufficient monitoring and complying, prioritization of cost-saving measures over others, targeting procedural harmonization instead of environmental performance, high rates of complexity, and exacerbation of North–South inequities between rich and poor country firms.

Individually, each of these weaknesses limits the capacity of these programs to mitigate GHG emissions by imposing a cost structure on these emissions at the firm level. As a consequence, the attempt by each of these initiatives to move from a hierarchical to horizontal monitoring structure faces some important obstacles. More broadly, these weaknesses create two important problems for the use of voluntary climate change programs. First, these weaknesses combine to create problems in both effectiveness and accountability. For example, without an adequate enforcement mechanism to ensure that firms account for their commitments in reducing, disclosing, standardizing or trading emissions, it is difficult to ensure the initiative is effective in targeting environmental performance and ultimately reducing emissions. Similarly, North–South inequities consistently limit the spread of important knowledge on emissions reductions strategy to firms in developing countries who will sooner contribute more to climate change than their Northern counterparts.

A second and more immediate consequence of these weaknesses for firms adopting private voluntary programs is a failure to generate the market conditions involved in regulated emission trading markets. Without these transparent market conditions, the financial sector and its vast resources remain immobilized in efforts to strengthen incentive structures for GHG mitigation within the real economy by raising capital and premium costs for firms that fail to reduce GHG emissions. Similarly, managers at the firm level fail to generate the necessary experience for remaining competitive in a GHG-constrained economy. Each of the initiatives in the typology constitutes important "design elements" common in many regulated emission trading markets that can potentially inform financial sector and investor decision-making. Carbon disclosure mechanisms, such as the CDP, replicate emissions trading registries, which are used to publicly disclose emission levels. Codes of conduct, such as the Action Plan, establish performance standards, which mirror the caps or baselines used in emissions trading to reduce emissions. EMS, such as ISO 14064 and 50001, provide sophisticated emissions counting and measuring methodologies, which are pivotal components in emissions trading monitoring and verification mechanisms. Finally, voluntary cap-and-trade and project offset markets replicate the supply and demand forces firms experience when involved in emissions trading.

In aggregate these programs not only fail to reduce emissions at the firm level, but they also create a complicated and patchwork market signal that the financial sector is unable to use as a basis for generating cross-sectoral change in a variety of industries by leveraging access to capital as an incentive to reduce GHG emissions. From this perspective, private voluntary climate change programs offer not only little environmental value, but little incentive for the financial sector to leverage its wealth as a source of cross-sectoral climate change mitigation. As a result of this assessment, scaling back state-based regulatory mechanisms in favor of voluntary corporate initiatives will do little to contain GHG emissions and costly climate change impacts.

Notes

1. Auld, Bernstein and Cashore (2008) also note the relative lack of categorization which leads to confusion when discussing voluntary initiatives. Though we agree that categorization needs to take place, we focus here on international-level initiatives, which results in somewhat different categorization from theirs.
2. Ibid. p. 7.

REFERENCES

Alliance to Save Energy. 2008. *Raising Standards: ISO Management System Standard for Energy.* Available at ⟨http://ase.org/content/article/detail/5466⟩; accessed April 23, 2009.

Auld, G., S. Bernstein, and B. Cashore. 2008. "The New Corporate Social Responsibility". *Annual Review of Environment and Resources* 33: 413–435.

Backstrand, K. 2006. "Democratizing Global Environmental Governance? Stakeholder Democracy after the World Summit on Sustainable Development." *European Journal of International Relations* 12 (4): 467–498.

BankTrack. 2008. *Meek Principles for a Tough Climate: Why the Carbon and Climate Principles Will Not Stop the Melting of Ice.* Available at ⟨http://www.banktrack.org/download/meek_principles_for_a_tough_climate/090329_meek_principles_for_a_tough_climate.pdf⟩; accessed April 24, 2009.

Boehmer, K., and A. Cherp. 2003. *ISO 14064: An Emerging Standard on Greenhouse Gas Accounting and Verification.* Available at ⟨http://web.ceu.hu/envsci/aleg/research/ISO-EnvFinance110503.pdf⟩; accessed April 23, 2009.

Brown, H. S., M. de Jong, and T. Lessidrenska. 2009. "The Rise of the Global Reporting Initiative: A Case of Institutional Entrepreneurship." *Environmental Politics* 18 (2): 182–200.

Bruno, K. (with J. Karliner). 2002. Greenwash + 10: *The UN's Global Compact, Corporate Accountability and the Johannesburg Earth Summit. CorpWatch.* ⟨http://s3.amazonaws.com/corpwatch.org/downloads/gw10.pdf⟩; accessed May 14, 2009.

Bumpus, A., and D. Liverman. 2008. "Accumulation by Decarbonization and the Governance of Carbon Offsets." *Economic Geography* 84 (2): 127–155.

Capoor, K., and P. Ambrosi. 2007. *State and Trends of the Carbon Market 2007. World Bank.* Available at ⟨http://carbonfinance.org/docs/Carbon_Trends_2007-_FINAL_-_May_2.pdf⟩; accessed April 24, 2009.

Carbon Finance. 2009. *Canadian Hedge Funds Seek Carbon Profits.* Available at ⟨http://www.carbon-financeonline.com/index.cfm?section=global&id=12036&action=view&return=home⟩; accessed November 20, 2009.

Carbon Principles. 2008. *The Carbon Principles.* http://carbonprinciples.org/documents/Carbon%20Principles.pdf; accessed May 14, 2009.

Cashore, B. 2002. "Legitimacy and the Privatization of Environmental Governance: How Non-State Market-Driven (NDSM) Governance Systems Gain Rule-Making Authority." *Governance* 15 (4): 503–529.

Chan, M. 2009. *Subprime Carbon? Re-Thinking the World's Largest New Derivatives Market.* Washington DC: Friends of the Earth.

Chatterjee, P., and M. Finger. 1994. *The Earth Brokers: Power, Politics and World Development.* London: Routledge.

Chicago Climate Exchange (CCX). 2009. *Emissions Reduction Commitment.* Available at ⟨http://www.chicagoclimatex.com/content.jsf?id=72⟩; accessed May 9, 2009.

Clapp, Jennifer. 1998. "The Privatization of Global Environmental Governance: ISO 14000 and the Developing World." *Global Governance* 4 (3): 295–316.

Clapp, Jennifer. 2005. "Global Environmental Governance for Corporate Responsibility and Accountability." *Global Environmental Politics* 5 (3): 23–34.
Clapp, Jennifer. 2008. "Illegal GMO Releases and Corporate Responsibility: Questioning the Effectiveness of Voluntary Measures." *Ecological Economics* 66 (2–3): 348–358.
Climate Disclosure Project (CDP5). 2007. *Carbon Disclosure Project Report 2007: Global 500.* ⟨http://www.cdproject.net/historic-reports.asp⟩; accessed May 14, 2009.
Climate Disclosure Project (CDP6). 2008. *Carbon Disclosure Project Report 2008: Global 500.* ⟨http://www.cdproject.net/download.asp?file=67_329_143_CDP%20Global%20500%20Report%202008.pdf⟩; accessed May 14, 2009.
Climate Disclosure Project (CDP 2009). 2009. *Carbon Disclosure Project Report 2009: Global 500.* ⟨https://www.cdproject.net/CDPResults/CDP_2009_Global_500_Report_with_Industry_Snapshots.pdf⟩; accessed November 20, 2009.
Climate Group. 2008. *The Climate Principles.* ⟨http://www.theclimateprinciples.org⟩; accessed May 14, 2009.
Esty, D. and A. Winston. 2006. *Green to Gold: How Smart Companies Use Environmental Strategy to Innovate, Create Value and Build Competitive Advantage.* New Haven CT: Yale University Press.
European Commission (EC). 2004. Commission Decision Establishing Guidelines for the Monitoring and Reporting of Green House Gas Emissions pursuant to Directive 2003/97/EC of the European Parliament and of the European Council. *EU Directive 2004/156/EC.* ⟨http://eur-lex.europa.eu/LexUriServ/site/en/oj/2004/l_059/l_05920040226en00010074.pdf⟩; accessed May 14, 2009.
European Commission (EC). 2008. EMAS and ISO/EN ISO 14001: Differences and Complementarities. *Eco-Management and Audit Scheme.* ⟨http://ec.europa.eu/environment/emas/pdf/factsheet/fs_iso_en.pdf⟩; accessed May 14, 2009.
Falkner, R. 2003. "Private Environmental Governance and International Relations: Exploring the Links." *Global Environmental Politics* 3 (2): 72–87.
Fuchs, D., A. Kalfagianni, and M. Arentsen. 2009. "Retail Power, Private Standards, and Sustainability in the Global Food System." In *Corporate Power in Global Agrifood Governance*, eds. J. Clapp and D. Fuchs, 29–49. Cambridge: MIT Press.
Gold Standard. 2008. *Gold Standard In Brief.* ⟨http://www.cdmgoldstandard.org/fileadmin/editors/files/1_communication/handout/GS_in_brief.pdf⟩; accessed May 3, 2009.
Gray, R., and J. Bebbington. 2001. *Accounting for the Environment.* London: Sage.
Gray, D., A. Murray, D. Sinclair, and D. Power. 2006. "Do Financial Markets Care About Environmental Disclosure? Further Evidence and Exploration from the UK." *Accounting, Auditing and Accountability* 19 (2): 228–255.
Gupta, A. 2008. "Transparency Under Scrutiny: Information Disclosure in Global Environmental Governance." *Global Environmental Politics* 8 (2): 1–7.
Hamilton, K., R. Bayon, G. Turner, and D. Higgins. 2007. *State of the Voluntary Carbon Market 2007: Picking Up Steam. New Carbon Finance and The Ecosystem Marketplace.* ⟨http://www.carbon.sref.info/an-example/market-news⟩; accessed April 30, 2009.

Hamilton, K., M. Sjardin, T. Marcello, and G. Xu. 2008. *Forging a Frontier: State of the Voluntary Carbon Market 2008. New Carbon Finance and The Ecosystem Marketplace.* ⟨http://www.carbon.sref.info/an-example/market-news⟩; accessed April 28, 2009.

Hebb, T. 2008. *No Small Change: Pension Funds and Corporate Engagement.* Ithaca NY: Cornell University Press.

Holme, R., and P. Watts. 2000. *Corporate Social Responsibility: Making Good Business Sense.* Geneva: World Business Council for Sustainable Development.

International Council on Mining and Minerals (ICMM). 2003. *10 Principles.* ⟨http://www.icmm.com/our-work/sustainable-development-framework/10-principles⟩; accessed May 14, 2009.

International Organization for Standardization (ISO). 2007. *The ISO Survey – 2007 shows ISO management systems standards implemented in 175 countries. ISO.* ⟨http://www.iso.org/iso/pressrelease.htm?refid=Ref1178⟩; accessed May 14, 2009.

Investor Network on Climate Risk (INCR). 2008a. *About INCR. Ceres.* ⟨http://www.incr.com/Page.aspx?pid=261⟩; accessed April 27, 2009.

Investor Network on Climate Risk (INCR). 2008b. *Investor Network on Climate Risk Action Plan. Ceres.* ⟨http://www.un.org/partnerships/Docs/Investor%20Network%20on%20Climate%20Risk%20Action%20Plan.pdf⟩; accessed May 14, 2009.

Islam, Md. S. 2008. "From Pond to Plate: Towards a Twin-Driven Commodity Chain in Bangladesh Shrimp Aquaculture." *Food Policy* 33 (3): 209–223.

ISO. 2011. Win the Energy Challenge with ISO 50001. International Organization for Standardization. ⟨http://www.iso.org/iso/iso_50001_energy.pdf⟩

Jagers, S., and J. Stripple. 2003. "Climate Governance Beyond the State." *Global Governance* 9 (3): 385–399.

Jagers, Sverker, Matthew Paterson, and Johannes Stripple. 2004. "Privatising Governance, Practising Triage: Securitization of Insurance Risks and the Politics of Global Warming." In *Business in International Environmental Politics: A Political Economy Approach,* eds. David Levy, Peter Newell, and Matthew Paterson. Cambridge, MA: MIT Press.

Kiernan, M. J. 2008. *Climate Change and Investment Risk.* Presentation at the Amsterdam Global Conference on Sustainability and Transparency, May 7–9, 2008. Amsterdam: Global Reporting Initiative.

King, A. A., and M. J. Lenox. 2000. "Industry Self-Regulation Without Sanctions: The Chemical Industry's Responsible Care Program." *Academy of Management Journal* 43 (4): 698–716.

Kolk, A., D. Levy, and J. Pinkse. 2008. "Corporate Responses in an Emerging Climate: Institutionalization and Commensuration of Carbon Disclosure." *European Accounting Review* 17 (4): 719–745.

Kollmuss, A., H. Zink, and C. Polycarp. 2008. *Making Sense of the Voluntary Market: A Comparison of Carbon Offset Standards. Stockholm Environment Institution and Tricorona.* ⟨http://www.sei-us.org/wwf_offset_standards_execsum.pdf⟩; accessed April 30, 2009.

Krut, R., and H. Gleckman. 1998. *ISO 14001: A Missed Opportunity for Global Sustainable Industrial Development.* London: Earthscan.

Lamberton, G. 2005. "Sustainability Accounting: A Brief History and Conceptual Framework." *Accounting Forum* 29 (1): 7–26.

Lenox, M. J., and J. Nash. 2003. "Industry Self-Regulation and Adverse Selection: A Comparison Across Four Trade Association Programs." *Business Strategy and the Environment* 12 (6): 343–356.

Lohmann, L. 2006. "Carbon Trading: A Critical Conversation on Climate Change, Privatisation and Power." *Development Dialogue* 48. Sturminster Newton, UK: Corner House.

Lohmann, L. 2008. "Carbon Trading: Solution or Obstacle?" In *The Impact of Climate Change on India,* 1–21. Sturminster Newton, UK: Corner House.

Missbach, A. 2004. "The Equator Principles: Drawing the Line for Socially Responsible Banks? An Interim Review from an NGO Perspective." *Development* 47 (3): 78–84.

Mesure, S. 2002. "Companies Told to Disclose Emissions." *The Independent*. June 1, 2002.

Nadvi, K., and F. Wältring. 2002. *Making Sense of Global Standards*. INEF Report. ⟨http://inef.uni-due.de/page/documents/Report58.pdf⟩; accessed May 14, 2009.

Nordhaus, W. D. 2007. "A Review of the Stern Review on the Economics of Climate Change." *Journal of Economic Literature* 45 (3): 686–702.

Paterson, Matthew. 2001. "Risky Business: Insurance Companies in Global Warming Politics." *Global Environmental Politics* 1 (4): 18–41.

Pattberg, P. 2005. "The Institutionalization of Private Governance: How Business and Nonprofit Organizations Agree on Transnational Rules." *Governance* 18 (4): 589–610.

Potoski, M., and A. Prakash. 2006. "Institutional Design for EMS-Based Government Procurement Policies." *Global Environmental Politics* 6 (4): 13–22.

Power, M. 1991. "Auditing and Environmental Expertise: Between Protest and Professionalization." *Accounting, Auditing and Accountability Journal* 4 (3): 30–42.

Prakash, A. 2000. "Responsible Care: An Assessment." *Business & Society* 39 (2): 183–209.

Prakash, A., and M. Potoski. 2006. *The Voluntary Environmentalists: Green Clubs, ISO 14001, and Voluntary Environmental Regulations*. Cambridge: Cambridge University Press.

Raynolds, L. 2000. "Re-embedding Global Agriculture: The International Organic and Fair Trade Movements." *Agriculture and Human Values* 17 (1): 297–309.

Raynolds, L. 2004. "The Globalization of Organic Agro-Food Networks." *World Development* 32 (5): 725–743.

Rutherford, P. 2003. "'Talking the Talk': Business Discourse at the World Summit on Sustainable Development." *Environmental Politics* 12 (2): 145–150.

Schmidheiny, S. 1992. *Changing Course: A Global Business Perspective on Development and the Environment*. Cambridge: MIT Press.

Sethi, S. P., and O. Emelianova. 2006. "A Failed Strategy of Using Voluntary Codes of Conduct by the Global Mining Industry." *Corporate Governance* 6 (3): 226–238.

Spencer, J., and C. Cui. 2007. "Green Thumb: The Climate for Investing, or Vice Versa." *Wall Street Journal*. September 22, 2007.

Therien, J. P., and V. Pouliot. 2006. "The Global Compact: Shifting the Politics of International Development?" *Global Governance* 12 (1): 55–75.

Toon, J. 2008. *Experts Working to Create a New International Standard for Energy Management. Renewable Energy World.* ⟨http://www.renewableenergyworld.com/rea/news/article/2008/12/experts-working-to-create-a-new-international-standard-for-energy-management-54189⟩; accessed May 14, 2009.

United Nations Conference on Trade and Development (UNCTAD). 2006. *World Investment Report 2006 – FDI from Developing and Transition Economies: Implications for Development.* New York and Geneva: United Nations.

United Nations Environment Program (UNEP FI). 2011. "About UNEP FI: Background," April 21, 2011. ⟨http://www.unepfi.org/about/background/index.html⟩.

United Nations Global Compact. 2008a. *Overview of the UN Global Compact.* ⟨http://www.unglobalcompact.org/aboutTheGC/index.html⟩; accessed May 14, 2009.

United Nations Global Compact. 2008b. *UN Global Compact: Note on Integrity Measures.* ⟨http://www.unglobalcompact.org/aboutTheGC/integrity.html⟩; accessed May 14, 2009.

United States Government Accountability Office. 2008. *International Climate Change Programs: Lessons Learned from the European Union's Emission Trading Scheme and the Kyoto Protocol's Clean Development Mechanism.* ⟨http://www.gao.gov/new.items/d09151.pdf⟩; accessed May 14, 2009.

United States House of Representatives. 2009. *Clean Energy and Safety Act.* House of Representatives. 111th Congress, 1st Session. March 31, 2009.

Utting, P. 2005. *Rethinking Business Regulation: From Self-Regulation to Social Control.* UNRISD Paper No. 15. Geneva: United Nations Research Institute for Sustainable Development.

Utting, P. 2008. "The Struggle for Corporate Accountability." *Development and Change* 39 (6): 959–975.

Utting, P., and J. Clapp, eds. 2008. *Corporate Accountability and Sustainable Development.* Delhi: Oxford University Press.

Vogel, D. 2008. "Private Global Business Regulation." *Annual Review of Political Science* 11: 261–282.

Wintergreen, J. 2007. ISO 14064 International Standard for GHG Emissions Inventories and Verification. Presented at 16th Annual International Emissions Inventory Conference. Raleigh NC: April 26, 2007.

Wright, C., and A. Rwabizambuga. 2006. "Institutional Pressures, Corporate Reputation, and Voluntary Codes of Conduct: An Examination of the Equator Principles". *Business and Society Review* 111 (1): 87–117.

Zadek, S. 2001. *The Civil Corporation: The New Economy of Corporate Citizenship.* London: Earthscan.

3

Private voluntary programs on climate change: US federal government as the sponsoring actor

Lily Hsueh and Aseem Prakash

Introduction

Private voluntary programs are rule structures that impose obligations on their participants that are beyond the legal requirements or where the law is silent.[1] These "beyond-compliance" obligations are expected to induce the program participants (or joiners) to produce positive social externalities. An externality implies that actors producing it do not fully internalize its costs and benefits. Consequently, actors tend to overproduce goods with negative externalities and undersupply goods with positive externalities. Reducing the emission of greenhouse gases (GHG) protects a global common and therefore creates positive externalities (Dolsak and Ostrom 2003). To increase the supply of positive externalities, a firm might adopt new technologies or management systems. But how will the firm be compensated for doing so, given that the benefits of its action are "external" to the firm? The policy challenge is to design institutions which encourage the production of positive externalities and, at the same time, reward the actors for doing so. Voluntary programs can be viewed as an institutional mechanism to pursue these dual objectives. Because voluntary programs serve this role with varying levels of efficacy, they need to be viewed with some level of healthy skepticism. With this in mind, the challenge for policy scholars is to outline an analytical framework which can help identify *ex ante* the externality-generating potential of programs. By doing so, policy scholars and analysts can assess *ex ante* the usefulness of a given program.

Business and climate policy: The potentials and pitfalls of private voluntary programs, Ronit (ed.), United Nations University Press, 2012, ISBN 978-92-808-1214-5

This chapter focuses on private voluntary programs in the climate change area. As Clapp and Thistlethwaite point out in Chapter 2, voluntary programs in the international sphere include a variety of initiatives such as broad proclamations, principles or codes of conduct, reporting mechanisms, environmental management systems and voluntary environmental markets. We are most interested in program design issues and therefore draw on the club approach, which provides an institutionalist perspective to understand the design of private voluntary programs across a variety of sectors. This approach suggests that private voluntary programs be viewed as "clubs" (Potoski and Prakash 2005b; Prakash and Potoski 2006a). In the economics literature, the central purpose of clubs is the production of club goods, which are non-rival but excludable, such as toll roads, sports facilities, theaters and museums (Buchanan 1965; Cornes and Sandler 1996; Tiebout 1956). In this perspective, it is not necessary for governments to provision non-rivalrous goods that tend to be used jointly by several citizens. As long as the supplier of goods can impose excludability (that is, allow only those who are paying to use or consume the good), such goods can be supplied by the market. Clubs thus emerge as an institutional mechanism to supply non-rival but excludable goods.

As Chapter 1 suggests, along with civil society organizations, business interest associations have been in the forefront of establishing programs to deal with the issue of reputations which are held in common by their members. In terms of sectors, private voluntary programs have proliferated in the environmental policy field (*Policy Studies Journal*, Special Issue 2007; Prakash and Potoski 2011). Interestingly, along with business interest associations, governmental bodies have also been in the forefront of sponsoring voluntary programs, especially in the environmental field. Among the national-level environmental regulators, the US Environmental Protection Agency (EPA) has been a leader in this aspect by sponsoring over 85 programs (EPA 2009a).[2]

Given the prior experience with this policy approach, it is not surprising that private voluntary programs have emerged as an important tool in US global climate change policy, especially during the Clinton and Bush II presidencies. In particular, given the opposition in the US Congress to climate change legislation, a policy space for creating private voluntary programs emerged which the EPA has exploited. In some ways, this also allowed the Bush Administration to shield itself from the criticism that the United States is standing on the sidelines regarding climate change. Given this unique political context, thousands of firms and organizations participate in private voluntary programs on climate change, sponsored by the federal government across a large number of industries and jurisdictions (EPA 2009b).

Many scholars have explored the institutional foundations of private voluntary programs to examine questions pertaining to their emergence, participation, design and efficacy (Coglianese and Nash 2001; Cutler, Haufler and Porter 1999; Potoski and Prakash 2009; Vogel 2005; Webb 2004). Indeed, Chapter 1 of this volume identifies four key aspects of this research: designing, joining, monitoring and complying. Prakash and Potoski's (2006a) club approach offers one way to think about institutional design, recruitment and efficacy issues. Firms might want to undertake beyond-compliance policies but are unsure about the excludable rewards they can capture for doing so. Outside stakeholders might want to reward firms that undertake beyond-compliance policies but may not have the means to identify the firms which have adopted such policies. The club approach suggests that private voluntary programs are a mechanism to fill this institutional gap. They facilitate the exchange of goodwill and reputational benefits from stakeholders in return for the beyond-compliance obligations which the participating firms agree to take on. In Vogel's (2005) terms, voluntary programs correct the institutional failures in the "market for virtue" and seek to align private with public goals.

More specifically, Prakash and Potoski (2006a) suggest that such programs (or clubs) can be viewed as institutional vehicles to achieve two purposes. Internally, these clubs induce participating firms to adopt beyond-compliance policies which lead to the production of desired externalities (e.g. reduction in GHG emissions). They do so by mitigating salient collective action issues that pertain to both recruitment to the program, and infringement of program obligations. Externally, these programs reduce information asymmetries between external stakeholders and firms regarding the internal policies participating actors have adopted, thereby empowering these stakeholders to reward or punish firms as they deem fit, a point that is also elaborated in Chapter 5 on programs in the agrifood sector by Fuchs and Boll. Thus, these programs serve as a branding vehicle which provides credible assurance regarding participating firms' commitment to produce positive social externalities. Some programs even create a certification logo to clearly communicate the brand identity.

The club approach as well as much of the voluntary program literature (Arora and Cason 1996; Barnett and King 2006; Potoski and Prakash 2009) posits that the provision of *excludable* reputational and goodwill benefits encourages firms to take on non-trivial costs of joining voluntary clubs. What types of excludable benefits will be provided is a function of several factors including club design.[3] Another key issue in club design is the provision of monitoring and enforcement mechanisms to ensure that members comply. The club approach recognizes that clubs vary in their capacities to generate positive externalities because clubs differ in their abilities to successfully recruit participants, and to ensure compliance with club obligations. In their design, effective programs therefore have rule structures to mitigate two central collective action problems: attracting firms to participate in the program (recruitment challenge) and ensuring that participating firms adhere to program obligations (shirking challenge). Because program efficacy can be undermined by collective action problems associated with free-riding and shirking, the effectiveness of voluntary clubs depends in part on how well clubs mitigate these challenges.

This chapter focuses on private voluntary clubs on climate change sponsored by the US federal government. Holding club sponsorship constant, the chapter seeks to understand the salient program features through the lens of club theory.[4] For our empirical analysis, we study four federal voluntary clubs – *WasteWise, Climate Leaders, Energy Star* and the *Voluntary Diesel Retrofit Program* – drawn from the US Environmental Protection Agency (EPA)'s database, which provides information on over 50 private voluntary programs on climate change sponsored by the federal government.[5]

The chapter is structured in the following way. In the first section, we review club theory as applied to voluntary programs. In particular, we theorize how voluntary programs are designed to incentivize firms to produce positive social externalities and to overcome institutional failures, which are inherent in the private provision of public goods. In the second section of the chapter, we employ the theory to explore the design of clubs sponsored by the US federal government. Finally, we conclude by identifying lessons for climate change policy.

Program design and the club framework

Clubs encourage actors to voluntarily incur private costs to produce positive social externalities. An externality implies that actors do not fully internalize the costs and benefits of their actions.[6] Government regulations are designed to compel firms to internalize negative externalities. This approach is an efficient mechanism if governments can correctly identify the source of the problem, supply regulations which can compel firms to reduce pollution, and have the capacity to monitor and enforce compliance. Because such assumptions about governmental capacities may not hold in every case, new regulatory tools have been proposed to supplement governmental regulation (Coglianese and Nash 2001; Fiorino 1999).[7] These include market-based instruments such as cap-and-trade systems (Tietenberg 2006), information-based regulation such as the Toxics Release Inventory Program (Konar and Cohen 1997) and voluntary clubs. It is important to appreciate that environmental governance is typically marked by institutional complexity whereby new regulatory approaches coexist with, and work under the shadow of, the more traditional public regulation, as Porter describes in relation to some programs in the automobile sector (Chapter 6).

Voluntary clubs can be viewed as correcting (albeit imperfectly) two types of institutional failures: market failures because markets do not incentivize regulatees to reduce pollution, and government failures because governments are unable to enact or enforce the laws designed to curb pollution. It is critical to recognize that voluntary clubs themselves are vulnerable to failure if they do not impose significant obligations on their participating members, or if they do not enforce compliance with club obligations. Indeed, some clubs may be designed to simply fool the public about the good intentions of the participating firms without imposing real obligations on participating firms. Such "greenwashes" are part of the voluntary program landscape. A good theory is therefore essential to help policymakers and scholars to differentiate *ex ante* the credible from the less credible voluntary programs so that stakeholders can focus their goodwill and attention on the former only.

Provision of excludable benefits is the key recruitment carrot that private voluntary programs can offer. Therefore, from a policy perspective, club sponsors must invest resources in communicating the club attributes and obligations to key stakeholders. Arguably, in the absence of such initiative, stakeholders may not be able to distinguish between effective voluntary clubs and greenwashes. Consequently, they may decide to treat all programs as greenwashes, and fail to reward any firm for its program participation. Such problems could lead to a "lemons market" (Akerlof 1970) for voluntary climate programs in which weak programs drive effective ones out of the market. What is important for clubs – and perhaps even central – is that they can build and communicate a brand identity that stakeholders understand and find credible.

In Prakash and Potoski's (2006a) approach, the central purpose of clubs is to induce the production of positive social externalities. These externalities can have the attributes of private goods (e.g. a voluntary club obligating participating firms to pay higher wages to indigenous coffee growers as in fair trade coffee), public goods (e.g. a voluntary club obligating participating firms to lower pollution as in the 33/50 program), common property resources (e.g. a voluntary club designed to protect fisheries as in the Marine Stewardship Council) or even club goods (e.g. a voluntary club obligating participating forestry firms not to cut trees which are revered by an aboriginal group, as in the Forest Stewardship Council).

Because the provision of positive externalities is expensive, clubs need to offset them by providing sufficient levels of non-rival but excludable benefits (i.e. benefits with characteristics of club goods) to the participants. Unlike the economics literature where clubs are defined by the characteristics of the goods they supply, in the Prakash and Potoski version, clubs are identified by the characteristics of the inducements they offer to their participants for supplying the desired externality.

In the club approach, the analytically salient benefit that the participants receive for producing externalities is the affiliation with the club's positive brand reputation, a non-rival but excludable benefit. In its broadest sense, voluntary club membership signals to firms' external stakeholders about members' environmental programs, policies and performance. In some contexts, this signal is viewed as valuable to both the participants and outside stakeholders because of its informational and assurance value. The assurance stems from the signal's collective and institutionalized character. If firms seek to signal such information on their own, it might lack credibility. However, when they signal it as a part of a broad collective whose rules they may not have set, it tends to be more credible.

Excludability of benefits is analytically important for club design. Without this key feature, shirking by participants could be a major problem that undermines a club's credibility. Instrumental actors may want to establish a reputation for environmental responsibility without having to incur the costs of taking on new obligations. They might hope that the goodwill created by environmentally responsible firms will spill over to them because the stakeholders, who cannot always identify which firms are doing the good deeds, will spread their rewards broadly. Effective clubs seek to solve such free-riding because they make excludable the benefits from producing positive externalities. They provide the necessary information to stakeholders, who can then target their rewards only to firms that have joined the club, and exclude non-participants. Thus, monitoring and enforcement mechanisms to enhance compliance are also key dimensions of program design but, of course, these must also be unfolded in practice once programs are established. The next section of the chapter discusses the challenges and opportunities club sponsors face in designing clubs that limit shirking of club obligations by members and free-riding by non-members.

Mitigating collective action dilemmas through institutional design

All institutions can (and do) fail. Governments fail. Markets fail. Voluntary clubs fail as well. Once we recognize that we live in an institutionally imperfect (or a failed!) world, it is important that in evaluating the success of any institutional type, we do not establish unrealistic benchmarks. This point was effectively made by Coase (1991) in his Nobel Prize acceptance speech.[8] Both the detractors (who tend to focus on club failures and ignore governmental failures) and the supporters (who tend to focus on governmental failures and ignore club failures) of voluntary clubs need to exercise caution in this regard. Unlike Coase, we do not suggest that clubs replace governmental regulation. We view programs in climate change policy as supplementing market processes as well as governmental regulation (Potoski and Prakash 2005a), especially in developing countries which are characterized by poorly functioning markets as well as governments which lack the capacities (and/or incentives sometimes) to enforce their own regulations (Dasgupta, Hettinge and Wheeler 2000), which is a particularly pertinent issue in current climate change policy. Clubs are bridges which help us to navigate through institutional failures, with the objective of providing more effective market and regulatory systems. From a policy perspective, the objective is to understand the conditions under which private voluntary programs fail and how their institutional design, as the key independent variable in their efficacy, can mitigate or contribute to their failure.

One type of free-riding pertains to shirking: firms can join a club and claim to produce positive social externalities but fail to live up to their promises. The club therefore needs to establish mechanisms to compel participants to adhere to program obligations. Once programs are created, however, widespread shirking can undermine the production of environmental externalities and thereby dilutes its credibility. Willful shirking occurs because: (1) the goals of participants and voluntary club sponsors diverge; and (2) participants are able to exploit information asymmetries (regarding their adherence to a program) between themselves and sponsors and stakeholders. Information asymmetries prevent stakeholders from differentiating program shirkers from non-shirkers.

Voluntary clubs can mitigate shirking by establishing monitoring and sanctioning mechanisms as part of the formalization of programs. A voluntary club with a reputation for effectively policing and sanctioning its participants is likely to have a stronger standing among its stakeholders and therefore have a stronger brand reputation among its firms' stakeholders.

The Olsonian free-riding dilemma: Brand benefits and club standards

With traditional public regulations as the baseline, club standards specify what beyond-compliance actions are required for firms to join a program and remain members in good standing. Some standards specify performance requirements (sometimes called outcome standards) while other standards may be more process oriented, such as requirements that members adopt a management system, or that members regularly consult with community groups. Finally, club standards may limit membership to those that have already established high standards of environmental performance. In effect, program rules are signals to members' stakeholders on what the voluntary club wants members to accomplish, particularly their production of environmental externalities. The stringency of standards serves as a proxy signal for the level of externalities members generate (per capita) and therefore affects the branding benefits members can expect to receive from stakeholders.

Moreover, the excludability feature of club standards is analytically important for club design because of the possibility of free-riding by those who want to enjoy club benefits without adhering to club obligations. Instrumental actors may want to establish a reputation for environmental responsibility without having to incur the costs of taking on new obligations. They might hope that the goodwill created by environmentally responsible firms will spill over to them. Thus, the provision of excludable benefits is critical for the success of the club. For external actors which seek to recognize or grant some sort of excludable benefits to club members, the stringency of club standards provides a sense of the social externalities clubs produce, and hence allows them to calibrate the level of excludable benefits they want to bestow on club members. Consequently, club standards mitigate the Olsonian free-riding or recruitment challenge (Olson 1965), which is the first source of institutional failure for voluntary clubs.

While voluntary clubs establish regulations outside the scope of mandatory government law, or when the law is silent, it is through reference to the requirements of mandatory government regulations that we can more precisely observe the degree of the "voluntary" component in voluntary clubs and assess the levels of externalities the programs produce. The voluntary nature of these programs stems from firms' behavior that produces "positive" social outcomes – positive social externalities – beyond what public law requires. This is relevant both in cases when programs are sponsored by actors external to the business community, as is the case in this chapter, and in areas where programs are brought to life by business itself. This means of course that the same action that is voluntary in a jurisdiction with less stringent public law (or where the public law is silent, as in the case of US climate change policy) could be mandatory in a jurisdiction with stringent public law.

Public law is also the analytic referent for measuring the policy contribution of a program to social welfare: how much more positive social externality does a program compel its members to produce than they would produce in the absence of the program? The marginal contribution to public welfare from a voluntary club is the value added from its participants' activities that are beyond the applicable legal requirements. Again, this means that a voluntary club may contribute to public welfare in a jurisdiction with less stringent public law but may offer little or no contribution in a jurisdiction with stringent public law. This is quite important in the case of the United States which has yet to sign the Kyoto Protocol or enact domestic legislation compelling firms to reduce emissions of GHGs, and in light of the lack of a clear international roadmap agreed upon by rich and developing nations on climate change despite the 2009 Copenhagen Summit on climate change.

To simplify our discussion, we identify two types of standards in private voluntary programs with different levels of resources and program administration. Lenient club standards require little social externality production from members beyond what government regulations require. These are low-cost voluntary clubs for the members but create marginal levels of social externalities, and therefore the value of their brand among stakeholders is relatively low. Of course, even lenient club standards must mandate that members produce some positive social externality, or else the voluntary club would be a mere empty gesture (as some voluntary clubs indeed are).

Stringent club standards require members to produce high levels of positive social externalities, well beyond what government regulations require. For potential participants, these can be high-cost clubs. The advantage of stringent standards is that the club's brand is more credible and serves as a relatively low-cost tool for signaling voluntary club members' commitment to the club's social objective. Stakeholders can easily and confidently distinguish leaders (members) from laggards (nonmembers) among firms. Armed with this information, stakeholders can reward and punish firms accordingly.

While it is theoretically simple to talk about standards in private voluntary programs, we recognize that some stakeholders may find it difficult to evaluate the extent to which specific standards generate positive externalities (how high is high, or how low is low). Sophisticated stakeholders, such as well-funded environmental groups, as documented in Chapter 4 by Bozzi, Cashore, Levin and McDermott, or government regulators as in our case, may be able to interpret a club's brand signals. Less sophisticated stakeholders such as ordinary consumers may need some assistance in translating the brand signal into useful information for guiding their purchases. They may take cues from established actors such as civil society organizations which are known for their technical expertise. Some may seek other types of information shortcuts, such as the attributes of the sponsor, to evaluate a club's brand signal. And some may simply ignore it: arguably, just because new information about a company or a product is made available, it is not necessary that consumers will systematically embed it in their decision-making processes (Weil, Fung, Graham and Fagotto 2006). Thus, it is important that the club sponsors invest resources to clearly communicate the extent to which clubs impose beyond-compliance obligations on their participating members. In particular, they need to target this communication to key opinion leaders. In some cases, they may even wish to involve some key opinion leaders in the formulation of the club design, as has been documented in the case of the forestry industry (Cashore, Auld and Newsom 2004).[9] In some ways, the credibility of this signal is also influenced by the credibility of the sponsoring actor. Two sponsors might formulate clubs which impose similar obligations. However, stakeholders may assess the credibility of these clubs in a different manner, in response to varying reputations of the club sponsors (Darnall, Potoski and Prakash 2009).

The shirking dilemma: Monitoring and enforcement rules

While free-riding by non-members is the first source of institutional failure, shirking by club members is the second source of institutional failure for voluntary clubs. Shirking implies that some participants formally join the club but do not implement and practice the club standards. In doing so, shirkers seek to free-ride on the efforts of other members who build the voluntary club's reputation. While non-members are excluded from enjoying the benefits of club membership, shirkers enjoy club benefits unless they are discovered and expelled from the voluntary club. As word spreads about large-scale shirking, the program's reputation is likely to diminish and the brand reputation be undermined.

Willful shirking is facilitated by information asymmetries between voluntary club participants and club sponsors and/or between participants and club stakeholders. By information asymmetries we mean that voluntary club sponsors and stakeholders cannot observe the levels to which an individual participant is adhering to club standards because such activities are inherently difficult to monitor effectively or are observable only at significant cost. The net effect is that information asymmetries impose costs on sponsors and stakeholders seeking to differentiate program shirkers from non-shirkers.

Shirking violates appropriate behavior norms (March and Olsen 1989), which suggests that shirking can be curbed by social pressures (normative, mimetic and coercive) from other participating firms or even stakeholders. It may be important to understand the general conditions under which such social pressures would persuade instrumental firms not to shirk (Rees 1997). As scholars interested in studying the consequences of institutional design on collective action, we are more interested in studying how institutional design can address the issue of shirking.

Instead of relying on social pressures alone, a program might seek to mitigate shirking through its institutional design. Monitoring and enforcement mechanisms can compel members to adhere to club standards, particularly if they contain three central components: third-party monitoring, which could be government agencies, public disclosure of audit information, and sanctioning by program sponsors. When voluntary environmental clubs have none of these components – as in the Sustainable Slopes Program (Rivera and deLeon 2004), they tend to function as greenwashes. Based on the design features, we expect such programs to exhibit high levels of shirking and therefore generate very small amounts of positive externalities, if any. Indeed, Rivera and deLeon report that Sustainable Slopes club participants were no greener than non-participants. Our framework suggests that policymakers and stakeholders should be skeptical of clubs without any monitoring and enforcement rules, as these are key elements in the operation of any private voluntary program. After all, as Hobbes (1651) correctly observed, "Covenants without swords are but words, and of no strength to secure a man at all."

Private voluntary programs begin to have some credibility regarding their capacity to curb shirking if they exhibit at least one of the three features. Third-party monitoring means that firms are required by the program sponsor to have their policies audited by accredited, external auditors. Thus, the program might stipulate that a periodic approval granted by a third-party auditor is necessary to retain program membership. In some cases, program sponsors may require public disclosure of audit information (as in the European Union's Eco-Management and Audit Scheme). The idea is that, by such disclosure, the stakeholders can reward and punish as they deem fit. Finally, the sponsoring organization may itself act upon the audit information and sanction the shirkers.

We characterize a program's monitoring and enforcement programs as "swords", indicating that private voluntary programs may have different degrees of maturity. Strong/medium sword clubs have most of these components – audits, disclosure and sanctioning mechanisms – and are most likely to curb shirking because they provide for a monitoring mechanism, mitigate information asymmetries between participants and club sponsors/stakeholders, and create mechanisms for sponsors to sanction shirkers. In extreme cases, sponsors may expel participants from the program, an undesirable outcome for firms if they value the benefits of voluntary club membership. While strong/medium sword clubs should experience less shirking, they can impose more costs on members. Thus, in thinking about program design, policymakers need to examine the

Table 3.1 Analytical typology of voluntary clubs.

	Club standards	
Swords	Lenient	Stringent
Weak/ negligible	Social externalities: low Shirking: high Branding benefits: marginal Cost: low	Social externalities: low Shirking: high Branding benefits: marginal Cost: moderate–high
Strong/ medium	Social externalities: moderate Shirking: low Branding benefits: low–moderate Cost: low–moderate	Social externalities: high Shirking: low Branding benefits: high Cost: high

marginal addition to overall branding benefits by strengthening a club's swords.

Weak swords clubs have only some of three monitoring components while negligible sword clubs virtually have no mechanisms to monitor or enforce compliance. ISO 14001 is an example of a weak sword club. The International Organization for Standardization (ISO), the sponsoring organization, is not known to aggressively sanction the shirkers. Importantly, the absence of public disclosure of audit information weakens stakeholders' ability to sanction shirking. However, these are also low-cost voluntary clubs and therefore within the financial means of a larger number of firms, as witnessed by the more than 154,572 facilities across 148 countries that had joined the ISO 14001 club as of December 2007.

Based on the above discussion, we identify four types of private voluntary programs (Table 3.1). Important arenas for our discussion and for future research include: how does the institutional-stakeholder environment along with firm characteristics (the relative salience of leaders versus laggards in the population) influence the emergence of various voluntary club types; what is the aggregate impact of a voluntary club in terms of the production of positive environmental externalities, defined as the product of externalities produced by each firm and the total number of club participants. In some instance, policymakers might favor lenient-standard clubs to attract a large roster as opposed to stringent-standard clubs with limited membership. In other instances, lenient-standard clubs might be labeled as greenwashes and attract few members simply because they cannot generate significant branding benefits. Thus, instead of one-type-fits-all, policymakers should recognize that different private voluntary programs are likely to best fit different policy contexts for different types of firms. While stringent-standard clubs with strong swords might seem the best from an externality generation perspective, these are high-cost clubs that most firms might not find worth their while.

On the other hand, weak sword clubs with lenient standards might generate low levels of externalities per capita, but by attracting a large roster of firms, might lead to the generation of high levels of externalities in the aggregate.

Federal clubs in climate change policy: Empirical evidence

Our research design follows the comparative case study method. We present and compare federal clubs drawn from the EPA's database of private voluntary programs on climate change. While these programs are across diverse sectors and can be characterized as multiple-sector programs that are key emitters of GHGs, such as agriculture, buildings, power generation, transportation and waste management, a majority of the federal clubs in our study promote energy efficiency and clean and renewable energy sources. Federal sponsors of private voluntary programs include the EPA, the Department of Energy (DOE), the Department of Agriculture (USDA) and the Department of Transportation (DOT), with a majority of programs initiated by the EPA. In addition to recruiting members predominantly from the commercial sector, some clubs recruit members from non-profits and government entities, including tribal governments, as well as the citizenry.

Our unit of analysis is the institution, i.e. the program design of clubs. Given that we are focusing on programs sponsored by US federal agencies, the reputation of the sponsoring actor can be deemed to be comparable across clubs. In Table 3.2, we provide an inventory of federal clubs across the salient features identified in the club approach.

To what extent do climate change private voluntary programs cohere with the club theory outlined by Prakash and Potoski (2006a)? By and large, we find that institutional designs of the federal clubs surveyed in this chapter correspond well with the predictions of club theory. These clubs seek to attract members with the promise of non-rival but excludable benefits: the provision of public recognition for members is a key feature though some clubs hold the promise of monetary benefits as well. All but a small handful of clubs offer reputational and goodwill benefits to participating firms and organizations. They specify ways by which members could gain public recognition for adhering to club obligations. Often, these clubs tend to offer multiple tiers of environmental awards contingent on the level of club obligations fulfilled. Examples include the EPA's Urban Heat Island Reduction Initiative and Open Space Preservation Strategies for Promoting Smarter Growth and Environmental Protection, as well as the DOE's Clean Cities Program and FreedomCAR and Fuel Partnership.

Table 3.2 Federal voluntary clubs, reputable benefits, standards and swords.[1]

Voluntary clubs	Federal agency sponsor	Sector	Standards (lenient, stringent)	Sword (no sword, weak, medium, strong)
AgStar	EPA; DOE; USDA	Agriculture	Stringent	Medium
Best Workplace for Commuters Program	EPA; DOT	Transportation	Stringent	Medium
BestPractices Program	DOE	Industrial	Stringent	Medium
Building Energy Codes Program	DOE	Buildings	Lenient	No sword
Building Technologies Program	DOE	Buildings	Stringent	Medium
Buildings Cooling Heating and Power (BCHP) Initiative	DOE	Buildings	Stringent	Medium
Carbon Sequestration Leadership Forum	DOE, DOS	Power generation/energy	Stringent	Medium
Clean Cities Program	DOE	Transportation	Stringent	No sword
Clean School Bus USA	EPA	Transportation	Stringent	Weak
Climate Leaders	EPA	Multisector	Stringent	Weak
Climate VISION	DOE	Industrial	Stringent	Weak
Coal Combustion Products Program (C2P2)	FHWA, EPA, DOE, ACAA, USDA	Multisector	Lenient	No sword
Coalbed Methane Outreach Program (CMOP)	EPA	Industrial	Lenient	No sword
Combined Heat and Power (CHP) Partnership	EPA	Industrial	Stringent	Medium
Community Based Environmental Protection (CBEP)	EPA	Community development and smart growth	Lenient	No sword
Cooperative Conservation Initiative	DOI	Multisector	Stringent	No sword
Emerging Technologies Program	DOE	Buildings	Stringent	Medium
Emission Reduction Partnership for the Semiconductor Industry	EPA	Industrial/ semiconductors	Stringent	Medium
ENERGY STAR	EPA; DOE	Multisector	Stringent	Medium
EnergySmart Schools Program	DOE	Buildings	Stringent	Medium
Federal Energy Management Program (FEMP)	DOE	Buildings	Lenient	Medium

FreedomCAR and Fuel Partnership	DOE	Transportation	Stringent	Medium
Geothermal Technologies Program	DOE	Power generation/energy	Stringent	Medium
Green Communities Program	EPA	Community development and smart growth	Lenient	No sword
Green Engineering Program	EPA	Multisector	Lenient	No sword
Green Power Partnership	EPA	Power generation	Stringent	Weak
Green Suppliers Network (GSN)	EPA	Waste management	Lenient	Weak
GreenScapes Alliance	EPA	Multisector/land management	Lenient	No sword
Industrial Assessment Center (IAC) Program	DOE	Industrial	Stringent	Medium
Industries of the Future (IOF)	DOE	Industrial	Stringent	Medium
Integrated Environmental Strategies (IES)	EPA	Multisector	Lenient	Weak
It All Adds Up to Clean Air	DOT; EPA	Transportation	Lenient	No sword
Inventions and Innovation program	DOE	Multisector	Stringent	Weak
Landfill Methane Outreach Program (LMOP)	EPA	Waste management	Stringent	Medium
Methane to Markets (M2M)	CEQ; DOE; DOS; EPA; USAID; USTDA	Multisector	Stringent	Weak
Mobile Air Conditioning Climate Protection Partnership	EPA	Transportation	Lenient	No sword
Natural Gas Star	EPA	Industrial	Lenient	No sword
Open Space Preservation Strategies for Promoting Smarter Growth and Environmental Protection	EPA	Community development	Stringent	No sword
Plug-in to eCycling Program	EPA	Waste management	Stringent	Weak
Sector STAR	EPA	Multisector	Lenient	No sword
Sector Strategies Program	EPA	Multisector	Lenient	No sword
SF6 Emission Reduction Partnership for Electric Power Systems	EPA	Industrial/electric power	Stringent	No sword
SF6 Emission Reduction Partnership for the Magnesium Industry	EPA	Industrial	Stringent	No sword

Table 3.2 (cont.)

Voluntary clubs	Federal agency sponsor	Sector	Standards (lenient, stringent)	Sword (no sword, weak, medium, strong)
Smart Communities Network	DOE	Community development and smart growth	Lenient	No sword
Smart Growth Network (SGN)	EPA	Community development and smart growth	Lenient	No sword
SmartWay Transport	EPA	Transportation	Stringent	Medium
Solar Energy Technology Program	DOE	Power generation/energy	Stringent	Medium
Urban Heat Island Reduction Initiative	EPA	Multisector/building technologies; forest management	Lenient	No sword
Voluntary Aluminum Industry Partnership (VAIP)	EPA	Industrial/aluminum industry	Lenient	No sword
Voluntary Diesel Retrofit Program	EPA	Transportation	Stringent	Strong
Voluntary Reporting of Greenhouse Gases Program	DOE	Multisector	Stringent	Weak
WaterSense	EPA	Multisector	Stringent	Strong
WasteWise	EPA	Waste management	Lenient	Weak
Weatherization Assistance Program (WAP)	DOE	Buildings	Lenient	Weak
Wind and Hydropower Technologies Program	DOE	Power generation/energy	Stringent	Medium

1. We exclude from our analysis voluntary programs that exclusively provide financial assistance; programs that are private–public partnerships whereby the government strategically targets specific firms or industry associations or regions for technology development; discontinued programs; programs that seek to build environmental regulatory capacity and infrastructure in developing countries; and programs that rely heavily on the states to deploy and implement projects. Excluded programs include: DOE's Stationary Power Fuel Cell Program, Hydrogen, Fuel Cells and Infrastructure Technologies Program, Distributed Energy and Electric Reliability (DEER) Program; USAID's U.S.–Asia Environmental Partnership (USAEP) State Environmental Initiative; EPA's Performance Track (discontinued); DOE's Rebuild America, DOE's Carbon Sequestration Regional Partnerships; Million Solar Roofs Program (discontinued); DOE's State Energy Program (SEP); DOE's Solid State Energy Conversion Alliance (SECA).

Table 3.3 Case selection.

	Club standards	
Swords	Lenient	Stringent
Weak/negligible	WasteWise	Climate Leaders
Strong/medium	N/A	Energy Star, Voluntary Diesel Retrofit Program

While the federal climate change programs vary in the stringency of their standards, they tend to have weak mechanisms for monitoring and sanctioning shirking by members. We, therefore, find that most federal clubs are, at best, clubs with weak swords. Examples include the EPA's Clean School Bus USA and the DOE's Climate Vision. Some clubs have no explicit swords; their chief purpose appears to be to promote environmental stewardship among consumers, firms and other organizations through information dissemination, with awards of excellence for participants who take extraordinary steps in meeting club obligations. Examples include the EPA's GreenScapes Alliance and the DOE's Clean Cities Program. With that said, several elite federal clubs possess medium swords (e.g. Energy Star, Carbon Sequestration Leadership Forum). Nationally acclaimed, these clubs are perceived to enjoy a great deal of public legitimacy. Very few federal clubs possess strong swords. One such club is the Voluntary Diesel Retrofit Program which we discuss below.

To explore institutional design issues in more detail, we focus on four programs, namely, WasteWise, Climate Leaders, Energy Star and the Voluntary Diesel Retrofit Program. As indicated in Table 3.3, we selected the cases to ensure variation in the two institutional design features, namely, club standards and monitoring and enforcement mechanisms to enhance compliance, while reflecting as much as possible the distribution of clubs in the EPA database.[10]

With respect to club standards, WasteWise exhibits lenient standards while Climate Leaders, Energy Star and the Voluntary Diesel Retrofit Program possess stringent standards. The Voluntary Diesel Retrofit Program and Energy Star possess strong/medium swords while Climate Leaders possesses a weak sword. WasteWise possesses negligible swords. Overall, WasteWise is a lenient-standard club with no sword. Climate Leaders is a stringent club with a weak sword. Energy Star has stringent standards and a medium-level sword. The Voluntary Diesel Retrofit Program is a stringent club with a strong sword. These combinations of standards and monitoring and enforcement mechanisms appear representative of the clubs in the EPA database. WasteWise, Climate Leaders and the Voluntary Diesel Retrofit Program are sponsored by the EPA, while Energy Star is jointly sponsored by the EPA and DOE.[11]

Through comparative case studies we illustrate how these clubs are designed to incentivize firms to produce positive social externalities and to solve collective action dilemmas – the Olsonian free-riding dilemma and the shirking dilemma – that are inherent in the private provision of public goods.[12] For each program, we first discuss member obligations and the club's stringency of standards. We next document the club's excludable benefits in terms of reputational and goodwill rewards. Finally, we designate the club's sword type, i.e. the club's monitoring and sanctioning mechanisms. We end each case study with a discussion of the club's efficacy with respect to club standards and monitoring and enforcement mechanisms.

WasteWise

WasteWise was launched by the EPA in 1994 to encourage firms, government and other public entities to reduce and recycle municipal solid wastes as well as selected industrial wastes by designing their own waste reduction programs tailored to their respective needs (EPA WasteWise 2010). On the face of it, WasteWise is not a "climate change" specific program. The reason it is listed on the EPA's website is that in reducing wastes, participating members tend to cut the emissions of CO_2, CH_4 and N_2O.

To become a member of WasteWise, firms must complete a baseline report and submit an annual report (which are not required by law) detailing the tonnage of waste reduced and associated cost savings. WasteWise assists participating firms and organizations to develop strategies for reducing wastes at the source. Because WasteWise membership obligations tend to complement rather than exceed the law, we categorize WasteWise's standards as lenient. This makes it a low-cost club. Not surprisingly, it has attracted a sizeable roster of participants. In May 2009, 2000 organizations participated in WasteWise, about 20 times more members than Climate Leaders (see next section).

WasteWise members go beyond compliance in terms of reporting requirements. There is some evidence that inventorizing waste and pollution is an important first step to generate momentum to reduce them (Prakash 2000). This is because most managers within firms tend not be cognizant of the levels of waste their firms generate. If they were to know this, they are more likely to take action to reduce it. The assumption is that inaction on the part of firms regarding waste reduction is rooted to some degree in ignorance instead of a strategic choice. Because WasteWise members are obligated to generate data that document waste generation, this is expected to encourage them to reduce their firms' carbon footprint.

Yet joining WasteWise entails some costs for which firms need some offsetting inducements. WasteWise has created a tiered system of public recognition. Along with their annual reports, participating firms are encouraged to submit a narrative of the steps they have taken to meet the waste reduction goals. WasteWise offers four tiers of awards annually in categories that include business, government and educational sectors. The WasteWise Hall of Fame is the highest honor awarded to members and indicates a continued commitment to waste reduction. The Partner of the Year Award recognizes overall waste reduction achievements, efforts to purchase or manufacture recycled-content products, and activities to promote the WasteWise program. There are two awards in the third tier: the Endorser of the Year, which recognizes organizations for outstanding efforts to promote the WasteWise program, and the Gold Achievement Award, which recognizes members' achievements in areas such as climate change, green buildings, product stewardship and the use of recycled materials. The last tier of public recognition is the Honorable Mention, which is awarded to members who have made good progress in waste prevention, recycling and the purchase of recycled products. In addition to these awards, WasteWise publishes success stories of members who have made significant progress toward their waste reduction goals.

WasteWise's lenient standards are coupled with little monitoring and enforcement, and no sanctions for non-reporting. An interview with WasteWise's staff reveals that compliance rate among members is low:[13] it was observed that "about 300 (or 15%) of our 2000 members submit their annual reports. It has been difficult to get firms to submit their reports. Contacting companies directly has not been very helpful because we end up talking to staff that do not have direct decision-making authority regarding their company's participation in the WasteWise program."[14] In sum, WasteWise has no sword: there are no formal requirements for third-party audit and public disclosure. Moreover, there is no sanctioning mechanism.[15] Shirkers are neither expelled from the club nor reprimanded in any other way.

One might suspect that WasteWise is not much more than a greenwash. It is a lenient club with no sword. However, there is one modest ray of hope. Because WasteWise is a low-cost club, it attracts a relatively large roster of firms. This, in some cases, might lead to the generation of high levels of externalities in the aggregate even if the per capita generation of positive externalities is low. We doubt this holds for WasteWise, an issue we hope scholars will empirically resolve. In sum, as a greenwash, WasteWise seems to have a limited potential in the mitigation of global climate change.

Climate Leaders

In the absence of federal regulation to limit the emissions of GHGs,[16] the EPA initiated Climate Leaders in 2002. This program is designed as a cooperation between industry and government to encourage firms across multiple sectors (e.g. semiconductors, chemical manufacturing, automobile, real estate) to develop long-term, comprehensive climate change strategies (EPA Climate Leaders 2010a). As of May 2009, Climate Leaders has 272 members, a much smaller membership roster than WasteWise, in part, because members are required to fulfill stricter obligations. Members are obligated to set corporate-wide GHG reduction goals and inventory their GHG emissions with the objective to measure progress toward their established goals. These club obligations are beyond what government regulations require.[17] Overall, we assess that Climate Leaders provides for relatively stringent club standards.

For potential joiners, membership obligations include the documentation of the emissions of six major GHGs (CO_2, CH_4, N_2O, HFCs, PFCs and SF_6) on a company-wide basis (including at least all domestic facilities) associated with onsite activities (fuel consumption and energy use, industrial process-related emissions, waste disposal, air conditioning/refrigeration use) as well as indirect emissions from electricity/steam purchases, and mobile sources. In emissions accounting, members must follow the inventory protocol developed by some of the new actors in climate change politics, namely the World Resources Institute (WRI; ⟨www.wri.org⟩) and the World Business Council for Sustainable Development (WBCSD; ⟨www.wbcsd.org⟩).

In return for meeting club obligations, Climate Leaders recognizes members' "climate leadership" by periodically publishing in *Newsweek, Fortune* magazine and other business media an advertisement with a list of all participating firms to congratulate "these Climate Leaders for taking action to reduce their greenhouse gas emissions and carbon footprint" (EPA Climate Leaders 2010b). Moreover, the Climate Leaders website, akin to WasteWise, publishes case studies of how corporate-wide climate change policies have transformed business operations (EPA Climate Leaders 2010c).[18]

Climate Leaders' stipulation that members go beyond the law in documenting their GHG emissions underscores their relatively stringent standards. These obligations, however, are not coupled with strong monitoring and enforcement rules. In effect, Climate Leaders is a weak sword club. While members are obligated to inventory and annually report their GHG emissions, there is no requirement for third-party audit. With that said, the program manager of Climate Leaders reports that "while at this time there is no formal requirement for third-party audit, many of the larger companies (members) choose to have it".[19]

With respect to public disclosure, members' GHG reduction objectives are displayed on the Climate Leaders website. When firms meet their goals, their achievements are recognized in Climate Leaders' annual report. However, there are no formal mechanisms for sanctioning shirking by club members. The policy of trusting participating firms' claims without verification, we believe, is problematic. Without consequences for shirking, there is no reason to expect members to be necessarily any greener than non-members (King and Lenox 2000; Potoski and Prakash 2005b).

Climate Leaders is perceived to be a high-cost club. Its relatively stringent standards attract membership from a select group of environmental leaders across diverse industries. These firms are willing to incur the medium to high costs of fulfilling stringent club obligations, in part for the non-rival but excludable benefits of highly visible public recognition in elite business and financial journals and magazines. However, the fact that Climate Leaders is weak in monitoring and enforcement suggests that its credibility for producing externalities is less than optimal. These dynamics may change if the EPA is successful in establishing a greenhouse emission registry, akin to the Toxics Release Inventory Program. If this were to happen, outside stakeholders would be able to benchmark the performance of Climate Leaders participants in relation to the non-participants. We speculate that such outside scrutiny and benchmarking is likely to serve as an effective sword because of the fear of bad publicity. In some ways, the success of this private voluntary program is dependent on the regulatory context in which it functions. This suggests that, instead of playing up the public versus private dichotomy, scholars need to examine the institutional complementarities, linkages and interdependence between various forms of regulation.

Energy Star

Established in 1992, Energy Star is a voluntary private program with certification rules initiated by the EPA to promote energy-efficient products with the aim to reduce GHG emissions (EPA Energy Star 2010a). In 1996, the DOE agreed to work jointly with EPA to promote energy-efficient products using the Energy Star label. Computers and monitors, followed by printers, were the first certified products under Energy Star. It appears that the initial recruitment success of this club was largely due to President Clinton's Executive Order 12845, which required that microcomputers, monitors and printers purchased by federal agencies be Energy Star-compliant (Webber, Brown and Koomey 2000). Organized to cover multiple sectors, over 50 product categories including major appliances, office equipment, lighting, and home electronics currently use the Energy Star label. The EPA has also extended the label to cover new homes and commercial and industrial buildings. Thus, the brand communication is quite visible and well established. In some ways, the Energy Star logo succinctly and effectively communicates its key benefit to consumers at large. In doing so, it helps consumers to clearly differentiate club participants from non-participants, thereby enhancing the excludability of the reputational benefits. We speculate this branding strategy mitigates the Olsonian free-riding dilemma and incentivizes firms to join this club.

Membership in Energy Star is contingent upon meeting strict energy efficiency guidelines set by the two federal agencies, the EPA and the DOE. Energy Star specifies detailed eligibility criteria and partner commitments across product lines, such as appliances (e.g. battery charging systems), heating and cooling (e.g. boilers), office equipment (e.g. monitors) and commercial food services (e.g. commercial dishwashers), among others (EPA Energy Star 2010b). For example, to earn the Energy Star label, manufacturers of battery charging systems for power tools, small household appliances and personal care products like electric shavers must meet energy performance specifications: battery charging system must not exceed a maximum non-active energy ratio, which is based on the nominal battery voltage (Vb). The EPA provides a table of maximum ranges allowed. These guidelines typically go well beyond what government regulations require: "(they) raise the bar of energy efficiency ..., and (encourage) superior energy management practices in the commercial and industrial sectors" (Energy Star 2007 Annual Report, p. 56).

For potential participants, meeting these obligations comes at considerable cost. In addition to meeting the technical criteria that are product specific, Energy Star members are urged to voluntarily commit to marketing the Energy Star brand, revise the company purchasing or procurement specifications to include Energy Star, and provide a plan to the EPA on a quarterly basis that outlines specific measures to go beyond the program requirements and to increase availability of Energy Star certified products. Only when these technical and business marketing obligations are met are firms and organizations permitted to place the Energy Star label on their products. Indeed, Energy Star is often used by energy companies to encourage energy efficiency considerations among home buyers. For example, PG&E, the leading electricity/power company, has launched an Energy Star Qualified New Homes Program. Its website notes:

> Now you can enjoy all the comforts of a new home while saving hundreds of dollars on your annual energy costs. Just make sure your next home is an Energy Star New Home. Energy Star is the US Environmental Protection Agency's label for energy efficiency – the same one you'll find on energy-efficient appliances, lighting and electronics. PG&E works with builders to ensure Energy Star New Homes exceed California's rigorous energy efficiency standards by at least 15 per cent. Plus, the quality of each home is verified by an independent, third-party inspector. California has a broad range of climates and micro-climates. Each Energy Star New Home is specially designed for the

unique climate of the area where it is built. And they come in a variety of styles and prices, so chances are there's one just right for you. (PG&E 2010)

In addition, the Energy Star label enables participating companies to appropriate monetary benefits. Buyers of some of the Energy Star products can get federal tax credit; thus the Energy Star logo can help companies to realize additional sales (EPA Energy Star 2010c). Furthermore, Energy Star qualified manufacturers are eligible to apply for public recognition awards. Each year, the EPA and the DOE honor firms and organizations that contribute significantly to reducing GHG emissions through energy efficiency. There are four different levels of Energy Star awards: Sustained Excellence, Partners of the Year, Awards of Excellence (by product categories) and Special Recognition. These awards are presented each year at a highly publicized ceremony, and Energy Star award winners are featured in *Fortune* magazine (or another business journal) in a special advertising section paid for by the EPA and the DOE (EPA Energy Star 2010d).

Energy Star's relatively stringent rules are coupled with medium monitoring and enforcement mechanisms. Energy Star is a stringent-standard club with a medium sword. Across product lines, participating firms and organizations must conduct product tests following the EPA and the DOE's procedures and protocols to demonstrate that they meet Energy Star's specifications. While these protocols are strict, the tests are conducted by prospective members themselves rather than by a third-party certifier. Manufacturers self-certify and then submit the qualified product data to the EPA on a semi-annual basis, while the EPA and the DOE reserve the right to conduct market or field tests to verify compliance – which can be viewed as a substitute for third-party audits. Any departure from Energy Star's guidelines means that the Energy Star label would be stripped from the product in question, although in practice this has been rare.[20] Finally, while Energy Star members are not required to publicly disclose the specifics of their environmental performance, they must provide information to users (via their corporate website and user's manual) about energy-saving features and operating characteristics of Energy Star qualified products. Such information transparency provides consumers and other stakeholders a quantifiable indication of the value of the Energy Star label. Energy Star's mix of rigorous standards with medium enforcement rules and sanctioning mechanism suggests it is a stringent club with a medium sword.

Energy Star has earned an elite reputation among consumers, businesses and other stakeholders (Brown, Webber and Koomey 2002).[21] According to Interbrand, an international brand consulting firm, "Energy Star has grown into a well-recognized consumer brand, the result of well-crafted strategies, market-defined insights, and a perseverance to always improve on the past" (Energy Star 2007 Annual Report). While hard data are not yet available, it seems that Energy Star's institutional design has enabled it to overcome to a large extent the two institutional failures – Olsonian free-riding and shirking dilemmas – that confront private voluntary programs. We speculate that participating members will tend to generate moderate levels of per capita externalities. However, the aggregate impact might not be sizeable, given the relatively small roster of the club.

Voluntary Diesel Retrofit Program

The Voluntary Diesel Retrofit Program is part of a larger EPA campaign, the National Clean Diesel Campaign (NCDC), to promote diesel emission reduction strategies. NCDC includes regulatory programs to address new diesel engines as well as "innovative programs to address the millions of diesel engines already in use" (EPA National Clean Diesel Campaign 2010); the Voluntary Diesel Retrofit Program falls under the latter category of "innovative" programs. The purpose of the EPA's Voluntary Diesel Retrofit Program is to evaluate the emission reduction capabilities of a given diesel engine emission control technology. Manufacturers who pass the EPA's evaluation process are placed on a Verified Technology List, a highly exclusive list. Through this process, the EPA intends to "support their use in the market while providing customers with confidence that verified technologies will provide emission reductions (namely, CO_2, CH_4, N_2O) as listed" (EPA Diesel Retrofit Technology Verification 2010a).

To become a member of the retrofit club, manufacturers must undergo a thorough technical review of the technology in question as well as tightly controlled testing to quantify emission reductions. While the retrofit club is a non-regulatory program offered to companies, manufacturers are subject to similar stringency of standards as that of mandatory certification: members must pass the appropriate Federal Test Procedure (FTP) and fulfill other requirements in order to receive verification and then must comply with post-verification requirements, including (but not limited to) assembly line audits, in-use compliance and emission warranties. The Voluntary Diesel Retrofit Program is therefore a voluntary club with stringent standards.

There are two paths to technology verification in the Voluntary Diesel Retrofit Program. At the pre-verification meeting, EPA's Office of Transportation and Air Quality (OTAQ) determines whether a technology requires full verification or not; this depends on whether a technology is seeking verification for the first time or whether the technology is an existing verified product with upgraded parts. New technologies are subjected to full verification under the Environmental Technology Verification (ETV) Program. Technologies not subject to ETV test can be verified by the manufacturers themselves subject to EPA guidelines. For the ETV test path, manufacturers must complete specific procedural and quality control requirements essential to the ETV program. Following completion of the testing and the ETV quality control checks, the information is submitted by ETV directly to OTAQ for final OTAQ verification. With this said, manufacturers under either path must submit emissions testing data that fulfill the EPA's testing protocols.

In return for meeting club obligations which are expected to lead to the reduction of CO_2, CH_4, N_2O and particulate matters, Voluntary Diesel Retrofit Program recognizes members' verified technologies by placing them on the EPA's Verified Technology List. For each technology, the Verified Technology List describes the emission reduction performance of the technology, lists the engines for which the verified technology is compatible and for which State Implementation Plan (SIP) credit may potentially be received, and describes the specific conditions or operating criteria that must exist in the engine/vehicle/equipment combination for the verified technology to perform (e.g. the use of ultra-low sulfur diesel fuel, installation of back pressure sensors, or a minimum exhaust temperature for the vehicle or piece of equipment).[22]

EPA's Verified Technology List is essentially a federal seal of approval for the diesel engine emission control technology under verification. This is an honor conferred only on 20 technologies (less than 15 companies); it is easy to differentiate club members from non-members because of the stringent requirements that members must fulfill for their technologies to be placed on the list. Of note, like the Energy Star, firms can expect to earn direct monetary benefits from the voluntary club's brand.

The Voluntary Diesel Retrofit Program's stringent standards are coupled with strict monitoring and enforcement and sanctions for non-compliance. The Voluntary Diesel Retrofit Program possesses a strong sword: the club requires members to undergo post-verification audits and disclose the results. There are also explicit rules regarding non-compliance. After a technology/company has been placed on EPA's Verified Technology List, it must pass the in-use emissions testing. The in-use test feature of the Voluntary Diesel Retrofit Program's club design is critical to whether or not the technology will remain on EPA's Verified Technology List. Technologies which fail to prove acceptable in-use performance will result in loss of emission credits and removal from the Verified Technology List. In-use performance requirements and parameters as well as sanctioning rules regarding non-compliance are explicitly stated on the voluntary club's website (EPA Diesel Retrofit Technology

Verification 2010b). Should the EPA determine strong bias in testing, the EPA may require corrective action, i.e. additional in-use testing of non-represented engines.

Unlike most other federal voluntary clubs in the EPA database, the Voluntary Diesel Retrofit Program is a stringent club with a strong sword. As a program that requires members to fulfill strict obligations, engage in audits and disclosure, as well as risk expulsion for non-compliance, the Voluntary Diesel Retrofit Program is more likely than Climate Leaders, WasteWise and Energy Star to curb shirking because it has instituted a monitoring mechanism and has created a mechanism for sponsors to sanction shirkers. A consequence is high positive externalities per verified technology – each verified technology achieves considerable reductions in CO_2, CH_4 and N_2O emissions. EPA's Verified Technology List is thus an exceptionally credible brand. With that said, a stringent club with a strong sword imposes more costs on members than the other types of private voluntary programs discussed in this chapter. The many steps to obtaining verification and passing post-verification tests are anything but trivial for club members. The high cost of joining the Voluntary Diesel Retrofit Program may be why only a small handful of technologies/companies are currently listed on EPA's Verified Technology List.

Conclusion

During the Clinton and Bush II presidencies, private voluntary programs emerged as an important tool in US global climate change policy, particularly given the divergence in the US Congress over climate change legislation. A policy space for creating private voluntary programs has thus emerged which the US EPA as well as other federal agencies has exploited with the creation of over 85 private voluntary programs. These programs exist alongside those created by civil society organizations and business interest associations to induce firms and other actors in climate change mitigation actions. This chapter introduces club theory as a conceptual lens to examine the institutional design of US federally sponsored private voluntary programs in relation to climate change.

In the first section of the chapter, we theorized how private voluntary programs are designed to incentivize firms to produce positive social externalities and to overcome institutional failures, which are inherent in the private provision of public goods. In the rest of the chapter, case studies of private voluntary programs sponsored by the US federal government – *WasteWise, Climate Leaders*, *Energy Star,* and the *Voluntary Diesel Retrofit Program* – were presented to illustrate the conceptual framework.

Empirically, we find that that the chief purpose of these federal programs is the promotion of environmental stewardship through information dissemination. Most federal programs seek to attract members through the provision of excludable reputational and goodwill benefits. In some cases such as Energy Star and the Voluntary Diesel Retrofit Program, participating firms can expect to earn direct monetary benefits as well.

Federal clubs vary in the stringency of their standards (in the EPA database, we find roughly an equal salience of clubs with stringent standards and clubs with lenient standards). Furthermore, these programs tend to have moderate to weak mechanisms for monitoring and sanctioning: indeed, the majority of clubs we surveyed possess weak swords or no swords at all. Arguably, the fact that these programs are sponsored by federal regulators lessens the necessity of third-party monitoring or public disclosures. After all, no firm would risk getting in the bad books of the regulators by shirking its program obligations. Yet, as we observe in the case of WasteWise, some participants tend not to take program obligations seriously; arguably they do not find the threat of federal reprisal to be credible.

With that said, several elite federal clubs such as Energy Star possess medium swords. These clubs are well publicized and they seem to enjoy a great deal of goodwill. Finally, there are very few clubs that possess strong swords – the Voluntary Diesel Retrofit Program is one of them. While EPA's Verified Technology List is a highly credible brand, the small number of technologies/companies in the Voluntary Diesel Retrofit Program limits the club's reach with respect to the promotion of environmental stewardship.

Our chapter raises important question for climate change policies. As noted in Chapter 7, there is an oversupply of voluntary programs in these sectors. Indeed, firms across a multitude of diverse industries have engaged different actors located at different territorial levels in overlapping initiatives without any effort towards coordination or coherence. Arguably, this oversupply is a reflection of or a cause of an undersupply of public regulation. Thus, this volume raises intriguing questions regarding the relationship between voluntary programs and public regulation. As global climate change negotiations remain deadlocked and unilateral national-level regulatory efforts are lacking in many countries, private voluntary programs may serve to fill the regulatory void. Importantly, voluntary programs may offer businesses the opportunity to signal their commitment to combating global climate change even though commitments by their governments on this issue are lacking.

Unlike other chapters which focus primarily on programs sponsored by a variety of private actors, our chapter focuses on clubs that are sponsored by governmental actors and joined by business. In our assessment, federal climate change clubs are a mixed bag. Some of them, such as Energy Star and the Voluntary Diesel Retrofit Program, can play an important role in overall climate change policies. However, as they are currently structured, most other clubs really work at the margin, if they work at all. Yet, the usefulness of these voluntary private programs may become more visible if governmental regulations begin to make exacting demands on various actors to reduce GHG emissions. In the absence of a muscular climate change policy, however, these marginal clubs will confuse the policy landscape by creating an illusion of policy activity, a strategy that the Bush II Administration tended to adopt.

In sum, by themselves, private voluntary programs cannot be expected to bring about important behavioral or policy change of the magnitude required for global climate change policies. While the previous administrations may have launched these clubs to dampen the demand for public regulation, it remains to be seen whether the current administration will engage voluntary approaches as a complement to public law or swing to the other extreme by shunning voluntary approaches. Moreover, private voluntary programs have an important international role to play as rich and developing nations continue dialog about global collaborative solutions, which was reemphasized at the 2009 Copenhagen Summit on climate change. Properly designed, these institutional vehicles can serve as a useful supplement to public regulation.

Notes

1. We recognize that there are isolated cases where the obligations imposed by private voluntary programs are below the threshold identified by public regulation. Oregon, in the case of voluntary programs in the forestry sector, is sometimes suggested as an example. While we recognize such outliers, it is important that theoretical perspectives be assessed in terms of their usefulness for understanding the median case and the population it represents, not outliers. Because the beyond-compliance attribute is useful for understanding the vast majority of programs across sectors, we believe that this should be *the* defining characteristic of any private voluntary program.
2. A 2005 survey identified 87 voluntary programs administered by the EPA, up from 54 in 1999 and 28 in 1996 (US EPA 2005). In fiscal year 2006, voluntary programs comprised 1.6 per cent of EPA's operating budget. Source: ⟨http://www.epa.gov/partners/⟩; accessed March 25, 2009.
3. The literature systemically studies how the institutional context and participants' characteristics influence the perceived value of these benefits. For example, firms in highly polluting industries might put a high value on club membership to signal their commitment to environmental responsibility (Potoski and Prakash 2005a). Or firms may place a high value on membership in response to supply chain pressures (Prakash and Potoski 2006b, 2007).

4. In presenting their formal model of the club approach, Kotchen and van't Veld (2009) show that sponsorship across government, industry and environmentalists under perfect and imperfect monitoring depends on the social welfare "weight" placed on public good benefits. Hsueh and Prakash (2009) posit that the form in which a club design offers excludable benefits depends on the institutional context in which the club sponsors function. Clubs sponsored by US states tend to emphasize information and capital provision over the reputational benefits emphasized in federal clubs.
5. We obtained detailed information about these individual clubs from the program websites. Furthermore, we interviewed program managers to understand the rationale for designing climate change clubs in specific ways.
6. Coase (1960) attributes externality problems to the absence of property rights. He conceptualizes pollution as a property rights problem: if one were to allocate property rights to the atmosphere, pollution problems might be solved more efficiently by private bargaining, instead of governmental intervention.
7. Some scholars view them as substitutes – a view we do not subscribe to.
8. Coase pointed to the tendency to benchmark imperfect markets against perfect governments. Soros (1997), on the other hand, criticizes the neoliberal perspective for benchmarking well-functioning markets against imperfect governments.
9. More generally, see the literature on stakeholder management on this issue. Given that there are multiple stakeholders, some with conflicting preferences, the managerial challenge is to decide which stakeholder to respond to. Mitchell, Agle, and Wood's (1997) framework identifying the three criteria of power, legitimacy, and urgency is one useful to think about this issue.
10. Our case selection strategy is consistent with King, Keohane, and Verba (1994)'s advice that "the best 'intentional' design selects observations to ensure variation in the explanatory variable ... without regard to the values of the dependent variables" (p. 140). Of note, all but a small handful of clubs in the EPA database offer reputational benefits; as such, we do not seek variation along that dimension.
11. Of note, in Table 3.3 we have intentionally offered no example of a lenient club with a strong/medium sword. There is only one federal sponsored club in the EPA database (i.e., the Federal Environmental Management Program) which is characterized by a medium sword with lenient club standards. As such, we consider this club an outlier rather than a representative club in our analytical framework.
12. We have obtained secondary data sources from program websites and gathered primary data via telephone interviews with program managers. We emailed a recruitment letter to several clubs including Climate Leaders, WasteWise and Energy Star on April 21, 2009. Program managers at Climate Leaders and WasteWise responded affirmatively to our interview request. Phone interviews with these two programs took place on May 7 and May 13, 2009, respectively. As of March 2010, we have not heard back from Energy Star despite repeated contact via email.
13. Delmas and Keller (2005) examine factors that explain the low reporting rate in the WasteWise program. Via a survey of 106 members (out of 947 companies), they show that for the WasteWise program collaborative behavior varies according to the initial reasons that the organization joined the club. Organizations will disclose information about their environmental performance when this disclosure provides them with the benefits they were looking for when they entered the program. Organizations that join WasteWise to promote their relations with the EPA and learn about waste techniques are more likely to submit an annual report. By contrast, firms that join the program because it is free are more likely to free-ride. Delmas and Keller also find that collaborative behavior increases with the involvement of upper management and decreases for later entrants in the program.

14. Interview with Amy Lile and Terry Grist, staff analyst and program manager of *WasteWise*, respectively, on May 7, 2009. WasteWise's staff further shared that they "expect and hope" annual reporting by members will increase with the deployment of a new database free of charge to members that will allow companies to readily track and report their waste generation, reduction and recycling data online.
15. Arguably, WasteWise's absence of monitoring is in part due to scarce resources – there are only two full-time and one half-time staff devoted to WasteWise at the EPA as of May 2009. This is compared to around 50 staff devoted to EPA's Energy Star program.
16. In September 2010, the EPA announced that the Climate Leaders program will phase down the services it offers in 2011 as a result of the EPA's Mandatory Reporting Rules, which the Climate Leaders program had helped to inform. The Mandatory Reporting Rules require mandatory reporting of GHG emissions from large sources in the US. Suppliers of fossil fuels or industrial greenhouse gases, manufacturers of vehicles and engines, and facilities that emit 25,000 metric tons or more per year of GHG emissions are required to submit annual reports to the EPA. The gases covered by the proposed rule are carbon dioxide (CO_2), methane (CH_4), nitrous oxide (N_2O), hydrofluorocarbons (HFC), perfluorocarbons (PFC), sulfur hexafluoride (SF_6), and other fluorinated gases including nitrogen trifluoride (NF_3), and hydrofluorinated ethers (HFE).
17. EPA's decision to regulate large emitters of GHGs means that *Climate Leaders* members who are large emitters will be expected to comply with EPA's newly proposed rules. Monica Neukomm, Climate Leaders' program manager, suggests that large emitters who have been members of Climate Leaders are in a better position than nonmembers to meet the new regulation. Telephone interview with Monica Neukomm took place on May 13, 2009.
18. As part of Climate Leaders' phase down, the EPA has identified a consortium of three nonprofit organizations with which it will co-sponsor a new national awards program to recognize exemplary corporate, organizational, and individual leadership in response to climate change. The consortium includes: The Climate Registry (The Registry), the Center for Climate and Energy Solutions (C2ES), and the Association of Climate Change Officers (ACCO).
19. Interview with Monica Neukomm on May 13, 2009.
20. It is not clear whether this is because *all* participating firms and organizations have met Energy Star thresholds for energy efficiency or because Energy Star has more lax enforcement rules in practice than on paper.
21. According to Webber et al. (2000), compared to the annual energy consumption of standard new equipment, typical Energy Star labeled products save the following percentage in energy consumption (depending on product type): office equipment, 30–70%; consumer electronics, 20–40%; residential heating and cooling equipment, 10–30%; residential and commercial lighting fixtures, 70–90%; and appliances, 10–50%.
22. State Implementation Plan (SIP) is a US state government plan for complying with the federal Clean Air Act, administered by the EPA. The SIP consists of narrative, rules, technical documentation, and agreements that an individual state will use to clean up polluted areas.

REFERENCES

Akerlof, G. A. 1970. "The Market for 'Lemons': Quality Uncertainty and the Market Mechanism." *Quarterly Journal of Economics* 84 (3): 488–500.

Arora, S., and T. N. Cason. 1996. "Why do Firms Volunteer to Exceed Environmental Regulations? Understanding participation in EPA's 33/50 program." *Land Economics* 72 (4): 413–432.

Barnett, M. L., and A. King. 2006. *Good Fences Make Good Neighbors: An Institutional Explanation of Industry Self-Regulation*. Paper presented at the Academy of Management Best Paper Proceedings, Atlanta GA.

Brown, R., C. Webber, and J. G. Koomey. 2002. "Status and Future Direction of the Energy Star Program." *Energy* 27: 505–520.

Buchanan, J. M. 1965. "An Economic Theory of Clubs." *Economica* 32: 1–14.

Cashore, B., G. Auld, and D. Newsom. 2004. *Governing Through Markets*. New Haven CT: Yale University Press.

Coase, R. H. 1960. "The Problem of Social Cost." *Journal of Law and Economics* 3: 1–44.

Coase, R. H. 1991. *Nobel Prize Lecture*. ⟨http://www.nobel.se/economics/laureates/1991/coase-lecture.html⟩; accessed August 15, 2007.

Coglianese, C., and J. Nash, eds. 2001. *Regulating from the Inside*. Washington DC: Resources for the Future.

Cornes, R., and T. Sandler. [1986] 1996. *The Theory of Externalities, Public Goods, and Club Goods*. 2nd edition. Cambridge, UK: Cambridge University Press.

Cutler, C. A., V. Haufler, and T. Porter, eds. 1999. *Private Authority and International Affairs*. Albany: State University of New York Press.

Darnall, N., M. Potoski, and A. Prakash. 2009. "Sponsorship Matters: Assessing Business Participation in Government and Industry Sponsored Voluntary Environmental Programs." *Journal of Public Administration Research and Theory* 19 (4): 1–25.

Dasgupta, S., H. Hettige, and D. Wheeler. 2000. "What Improves Environmental Compliance? Evidence from Mexican Industry." *Journal of Environmental Economics and Management* 39: 39–66.

Delmas, M., and A. Keller. 2005. "Strategic Free Riding in Voluntary Programs: The Case of the US EPA Wastewise Program." *Policy Sciences* 38: 91–106.

Dolsak, N., and E. Ostrom, eds. 2003. *The Commons in the New Millennium*. Cambridge MA: MIT Press.

Energy Star. 2010a. ⟨http://www.energystar.gov/index.cfm?c=home.index⟩; accessed April 27, 2010.

Energy Star. 2010b. ⟨http://www.energystar.gov/index.cfm?c=product_specs.pt_product_specs⟩; accessed April 27, 2010.

Energy Star. 2010c. ⟨http://energystar.custhelp.com/cgi-bin/energystar.cfg/php/enduser/std_alp.php⟩; accessed April 27, 2010.

Energy Star. 2010d. ⟨http://www.energystar.gov/ia/products/downloads/Fortune_Supplement_2008.pdf⟩; accessed April 27, 2010.

EPA. 2009a. ⟨http://www.epa.gov/partners/⟩; accessed March 25, 2009.

EPA. 2009b. ⟨http://yosemite.epa.gov/gw/StatePolicyActions.nsf/webpages/VoluntaryPartnershipPrograms.html⟩; accessed March 25, 2009.

EPA Climate Leaders. 2010a. ⟨http://www.epa.gov/climateleaders/⟩; accessed April 27, 2010.

EPA Climate Leaders. 2010b. ⟨http://www.epa.gov/climateleaders/documents/partner-recogniton/psa_dec_2007.pdf⟩; accessed April 27, 2010.

EPA Climate Leaders. 2010c. ⟨http://www.epa.gov/climateleaders/casestudies/index.html⟩; accessed April 27, 2010.

EPA National Clean Diesel Campaign. 2010. ⟨http://www.epa.gov/otaq/diesel/index.htm⟩; accessed April 27, 2010.

EPA Diesel Retrofit Technology Verification. 2010a. ⟨http://www.epa.gov/otaq/retrofit/index.htm⟩; accessed April 27, 2010.

EPA Diesel Retrofit Technology Verification. 2010b. ⟨http://www.epa.gov/otaq/retrofit/iu-testing.htm⟩; accessed April 27, 2010.

EPA Waste Wise. 2010. ⟨http://www.epa.gov/epawaste/partnerships/wastewise/index.htm⟩; accessed April 27, 2010.

Fiorinio, D. J. 1999. "Rethinking Environmental Regulation." *Harvard Environmental Law Review* 23: 441–469.

Hobbes, T. 1651. *Leviathan*. ⟨http://oregonstate.edu/instruct/phl302/texts/hobbes/leviathan-c.html⟩; accessed March 2, 2007.

Hsueh, L., and A. Prakash. 2009. *Voluntary Programs in Global Climate Change Policy: A Club Theory Perspective*. Paper presented at the 2009 annual conference of the Midwest Political Science Association.

King, A., and M. Lenox. 2000. "Industry Self-regulation without Sanctions: The Chemical Industry's Responsible Care Program." *Academy of Management Journal* 43: 698–716.

King, G., R. Keohane, and S. Verba. 1994. *Designing Social Inquiry*. Princeton NJ: Princeton University Press.

Konar, S., and M. A. Cohen. 1997. "Information As Regulation: The Effect of Community Right to Know Laws on Toxic Emissions." *Journal of Environmental Economics and Management* 32 (1): 109–124.

Kotchen, M., and K. van't Veld. 2009. "An Economics Perspective on Treating Voluntary Programs as Clubs." In *Voluntary Programs: A Club Theory Perspective,* eds. M. Potoski and A. Prakash. Cambridge MA: MIT Press.

March, J., and J. Olsen. 1989. *Rediscovering Institutions*. New York: Free Press.

Mitchell, R. K., B. R. Agle, and D. J. Wood. 1997. "Toward a Theory of Stakeholder Identification and Salience: Defining the Principle of Who and What Really Counts." *Academy of Management Review* 22 (4): 853–886.

Olson, M. 1965. *The Logic of Collective Action*. Cambridge MA: Harvard University Press.

PG & E. 2010. ⟨http://www.pge.com/myhome/saveenergymoney/eebuyersguides/newhome/⟩; accessed April 27, 2010.

Policy Studies Journal. 2007. Special issue: "Voluntary Environmental Programs: A Symposium". 35 (4): 685–792.

Potoski, M., and A. Prakash. 2005a. "Green Clubs and Voluntary Governance: ISO 14001 and Firms' Regulatory Compliance." *American Journal of Political Science* 49 (2): 235–248.

Potoski, M., and A. Prakash. 2005b. "Covenants with Weak Swords: ISO 14001 and Firms' Environmental Performance." *Journal of Policy Analysis and Management* 24 (4): 745–769.

Potoski, M., and A. Prakash, eds. 2009. *Voluntary Programs: A Club Theory Perspective*. Cambridge MA: MIT Press.
Prakash, A. 2000. *Greening the Firm*. Cambridge, UK: Cambridge University Press.
Prakash, A., and M. Potoski. 2006a. *The Voluntary Environmentalists*. Cambridge, UK: Cambridge University Press.
Prakash, A., and M. Potoski. 2006b. "Racing to the Bottom? Globalization, Environmental Governance, and ISO 14001." *American Journal of Political Science* 50 (2): 347–361.
Prakash, A., and M. Potoski. 2007. "Investing Up: FDI and the Cross-National Diffusion of ISO 14001." *International Studies Quarterly* 51 (3): 723–744.
Prakash, A., and M. Potoski. 2011. "Voluntary Environmental Programs: A Comparative Perspective." *Journal of Policy Analysis and Management* 31 (1). Published online September 14, 2011.
Rees, J. 1997. "The Development of Communitarian Regulation in the Chemical Industry." *Law and Policy* 19 (4): 477–528.
Rivera, J., and P. deLeon. 2004. "Is Greener Whiter? The Sustainable Slopes Program and the Voluntary Environmental Performance of Western Ski Areas." *Policy Studies Journal* 32 (3): 417–437.
Soros, G. 1997. "The Capitalist Threat." *Atlantic Monthly* 279 (2): 45–58.
Tiebout, C. M. 1956. "A Pure Theory of Public Expenditure." *Journal of Political Economy* 64: 416–424.
Tietenberg, T. 2006. *Emissions Trading: Principles and Practice*. 2nd ed. Washington DC: Resources for the Future.
Vogel, D. 2005. *The Market for Virtue: The Potential and Limits of Corporate Social Responsibility*. Washington DC: Brookings Institution Press.
Webb, K. ed. 2004. *Voluntary Codes: Private Governance, the Public Interest and Innovation.* Ottawa: Carleton Research Unit for Innovation, Science and Environment.
Webber, C. A., R. E. Brown, and J. G. Koomey. 2000. "Saving Estimates for the Energy Star Voluntary Labeling Program." *Energy Policy* 28: 1137–1149.
Weil, D., A. Fung, M. Graham, and E. Fagotto. 2006. "The Effectiveness of Regulatory Disclosure Policies." *Journal of Policy Analysis and Management* 25 (1): 155–181.

Part III

Sectors

4

The role of private voluntary climate programs affecting forests: Assessing their direct and intersecting effects

Laura Bozzi, Benjamin Cashore, Kelly Levin and Constance McDermott

Introduction

Any effort to understand and assess the role of private voluntary programs addressing climate change must pay careful attention to those initiatives that either affect, or turn to, forest management and governance as a means to ameliorate climate change in and across sectors. Forests are important to climate mitigation strategies because they are increasingly recognized as both significant sources and sinks of carbon dioxide (CO_2), although their climate impact varies across boreal, temperate and tropical forests. The loss and degradation of tropical forests in particular contributes a significant share of human-caused emissions, and will continue to do so over the next 10–20 years.[1]

Such a focus yields observations critical to understanding the past, present and future potential of private voluntary programs. First, as outlined in Chapter 7 of this book, differences occur within and across specific initiatives related to their degrees of "privateness", "voluntariness" and "programness". Likewise, forest carbon initiatives differ according to commercial interest-initiated efforts that focus on profit-maximizing incentives or competitive market logics, and programs initiated by civil society organizations that emphasize environmental and social values. These distinctions are important, as we reveal below, since they help place in context the emergence of competing forest carbon certification programs.

Second, and partly as a result, the policy scope is highly variable, with some forest instruments focused solely on carbon reductions, while others

Business and climate policy: The potentials and pitfalls of private voluntary programs, Ronit (ed.), United Nations University Press, 2012, ISBN 978-92-808-1214-5

entail additional environmental and social requirements. As a result there is a continuum with, at one end, programs that focus exclusively on verifiable carbon emissions offsets such as the Voluntary Carbon Standard and, at the other end, those primarily concerned with adding environmental and community development commitments to carbon offset projects, such as Plan Vivo.

Third, and related, voluntary carbon-focused programs always interact with other forms of private and public institutions within and outside carbon concerns, as outlined in Chapter 1. Building on this framework, we uncover an important interaction between private forest carbon programs, such as forest and carbon certification schemes, and innovative intergovernmental initiatives to channel billions of dollars towards climate mitigation for reducing emissions from deforestation and degradation (REDD).

This means that it is important not only to understand the direct impacts of forest-focused private voluntary programs in ameliorating emission reductions (which, we show below, appear quite limited), but also to uncover their *intersecting* effects with other private and public policy interventions, and to assess their "learning effects" that may create understanding and support among stakeholders to facilitate the development of more authoritative policy intervention efforts.

We illustrate these points in the following analytical steps. First, we review scientific understandings of the relationship between forests and climate change, a critical step in making informed assessments of policies that attempt to link forest and climate governance. This review is especially important for private authority where enthusiasm for new innovations often obscures their inability to significantly slow forest loss and/or reduce global emissions, inadvertently leading a range of scholars and practitioners to identify "false positives" regarding their direct problem-solving potential.

Second, we detail an analytical framework that pays attention to intersections and evolutionary aspects of private voluntary programs. We apply this framework by first reviewing the emergence of private/voluntary forest carbon markets. We then review the emergence of intergovernmental forest carbon offset programs, which, we reveal, interact with voluntary forest carbon markets.

We then turn our attention to a range of private and/or voluntary third-party carbon monitoring and verification schemes that have emerged to provide assurances that carbon trading schemes are either doing what they claim, or that provide assurance of additional non-carbon benefits.

Finally, we use the above review to reflect more deeply on two types of important interactions that emerge from our review: stakeholder learning about carbon markets; and the "symbiotic" potential of public and private authority to achieve more enduring and legitimate global impacts in ways that are not possible for individual public or private interventions.

The science of forests and climate change

Scientific and multi-stakeholder learning provided the initial catalyst for the integration of forests into the global climate regime as a particular sector. For example, in the mid-1990s, the UN Food and Agriculture Organization (FAO) engaged in periodic monitoring of the state of the world's forests, and has since produced a series of systematic assessments of global progress on forest-related issues. FAO's efforts reveal that global forest area has shrunk an estimated 3 per cent between 1990 and 2005, amounting to an average global loss of 13 million hectares per year (FAO 2006). This forest loss, compounded by population growth, has resulted in dramatic decreases in per capita forest area, resulting in further conflict and competition among different priorities for forest use, from fuel wood to timber production, carbon storage, biodiversity protection and recreation, among others.

Annual forest loss figures must be disentangled, however, because of the unequal distribution and impacts of forest loss on biodiversity and forest ecosystems. Between 1990 and 2005, the largest net loss in forest area occurred in South America (mostly in secondary and degraded forests), followed by Africa (though largely outside the megadiverse tropical Congo forests). However, the largest percentages of forest area lost were in Central America and Southeast Asia (FAO 2007; Geist and Lambin 2002; Rudel, DeFries, Asner and Laurance 2009). Likewise forest "degradation", involving loss of canopy cover and loss of structural diversity, can be as or more significant than deforestation in terms of its carbon and biodiversity impacts.[2] While degraded forests may retain their previous land use designation, their potential for climate mitigation and contribution to ecosystem functions may differ dramatically from forests in their primary state (DeFries et al. 2007). Forest vulnerability to climate change is also higher in degraded and/or intensively managed forests, creating a negative feedback loop (Malhi et al. 2008). Concern about degradation is particularly relevant to the African continent, where the loss of carbon through degradation is greater than the loss from deforestation (Gaston, Brown, Lorenzini and Singh 1998).

An increasing number of analyses and meta-analyses reveal that the direct and indirect causes of forest loss are highly complex, and include such diverse factors as global consumer demand, unsustainable logging practices, agricultural and pastoral expansion, rural poverty and displacement, war and civil unrest, and conflicting and/or ineffective government institutions and policies (FAO 2006; Geist and Lambin 2002; Rudel et al. 2009). This complexity, including interactions across world regions and across commercial and subsistence sectors, serves to complicate amelioration efforts.

In spite of this complexity, the Intergovernmental Panel on Climate Change (IPCC) reasons that forest-focused initiatives could have the largest impact on climate mitigation in the immediate future (Nabuurs et al. 2007). This is because scientific research has found that slowing deforestation and forest degradation, increasing or maintaining forest cover, enhancing "site-level" carbon density, as well as maintaining carbon in off-site wood products, can significantly reduce atmospheric concentrations of CO_2.

As an indication of just how important forests are to carbon reduction, some estimate that four-fifths of the Kyoto Protocol's first commitment period target could have been met through the elimination of deforestation activities in Brazil and Indonesia alone (Santilli et al. 2005). These findings led environmental non-governmental organizations and some business interest associations such as the World Business Council for Sustainable Development to consider forest protection as a potential win-win-win solution that could link the interests of both businesses and civil society by creating economic incentives to reduce carbon emissions, while also achieving broader social and environmental objectives. Understanding whether, when and how such positive synergies may occur, whether business interests can be married with public goals, and what pitfalls to avoid, requires the development of an analytical framework sensitive to the distinctions above and capable of addressing a highly dynamic environment in which multiple interests cooperate and compete to shape problem definitions and strategic interventions.

Analytical framework

Given the double-edged nature of forestry in climate policy – that deforestation and forest degradation contribute significantly to the climate problem, but the maintenance or growth of forests offer a near-term mitigation solution – we argue that an analytical framework must pay attention to identifying the particular properties of this interaction and address "on the ground" forest changes in a specific locale, as well as direct, indirect and intersecting effects on global forests and climate as a whole.

1. *On-the-ground forest changes*. For the purposes of broad comparison and measurement, and in an effort to assess the range of public and private approaches to forests and carbon, we draw on Hamilton,

Sjardin, Peters-Stanley and Marcello (2010a) who divide forest carbon projects into three broad categories: afforestation/reforestation (A/R), reduced emissions from deforestation and forest degradation (REDD) and improved forest management (IFM). They report that A/R projects have historically accounted for 63 per cent of all (publicly regulated and voluntary) forest carbon credits transacted (likely because the CDM is limited to A/R projects in the land use sector), followed by REDD at 17 per cent and IFM at 13 per cent. There is a recent trend towards projects that span multiple categories, with combined A/R, REDD and IFM projects constituting 24 per cent of all projects in 2008. Attention to these different project types focused our analysis not only on carbon reductions per se, but on the broader negative and positive management impacts that might occur.

2. *Direct effects*. Any effort to understand the effects of private and voluntary governance on forests must not only be problem-focused, but must distinguish direct from indirect impacts. By direct impacts, we refer to assessing whether, when and how support for voluntary and private governance efforts results in direct changes in reducing emissions. These can – drawing on the above – be distinguished based on criteria including whether they result in emissions reductions that otherwise would not have occurred (additionality); and result in emissions reductions that have a discernible impact in achieving critical targets, such as not going above 450 parts per million (ppm) of greenhouse gases (GHG) in the atmosphere. A focus on such direct impacts is important for identifying, as we do below, that support for private programs governing carbon reductions in the forest sector is currently woefully inadequate for creating any significant reductions within the global forest sector, and arguably is unlikely to do so in the future.
3. *Indirect and intersecting effects*. However, a focus on direct impacts only would downplay potentially significant indirect effects that private authority exerts on intergovernmental processes and public policy efforts in three key respects. First, private authority may intersect with other forms of public and private authority to generate unanticipated results. As we show below, private certification programs are emerging to both verify whether carbon-focused programs are actually doing what they say, from designing to complying, as well as expanding policy scopes to ensure that voluntary or private carbon emissions program efforts do not come at the expense of other environmental and social goals. Second, and related, private authority may intersect with intergovernmental processes to produce results that are deemed more legitimate, and result in greater and more expansive impact, than single private or public interventions, on their own, could accomplish. This builds on previous work by Levin, Cashore and Koppell (2009),

who theorize on "symbiotic" forms of private authority in which the private mechanism supports and gives legitimacy to a government intervention. Such an approach could avoid the situation in which hard-won intergovernmental or domestic public policy agreements have to be revisited. Symbiotic interactions in which two private voluntary programs layer atop one another is suggested here to indicate a similar dynamic, in which the co-benefits of a forest carbon project are separately verified from the carbon offset. Third, the role of private programs in either fostering, or being influenced by, policy learning and ideas at different stages of their development is a critically important but understudied role. As we show below, while intergovernmental processes may be slow to initiate and design formal forestry offset programs, their deliberations about key requirements may have profound effects on how private and voluntary programs are themselves designed. In turn, private and voluntary programs provide important learning-by-doing about such questions such as how to create viable cap-and-trade markets, or how to provide standards that ensure meaningful reductions. Such learning may then diffuse back to public or intergovernmental processes that pave the way for more effective and efficient public policy.

The key here is that the often painful learning processes, which at times derail public policy initiatives, are experienced instead in the private sphere and are hence avoided by intergovernmental or state-based solutions. Likewise the proliferation of different types of private programs may permit public officials to treat them as experiments in which they are able to adopt the most promising approaches at a later date. Similar potentials for learning can be observed in other sectors, and as illustrated in Tony Porter's chapter on the car industry, where regional arrangements can learn from each other. While learning and diffusion of ideas and technologies among public and private stakeholders is difficult to precisely track and measure, the impacts of this pathway could potentially be quite long lasting and profound. It is for these reasons that we conclude our chapter by considering the role that private forest initiatives may have played, and may continue to play, in fostering global learning about governance of forests in human-induced climate change. In this way we add to Karsten Ronit's observation, in Chapter 1, that public institutions shape the conditions for private regulation, by noting how private programs may likewise play a critical role in shaping the institutional conditions for public regulation.

Overall, our analytical framework points to the need to assess impacts from an evolutionary and intersecting perspective, and to consider the role of multiple stakeholders. While it is beyond the scope of this chapter to provide a full-fledged theory, we draw on this framework to review key developments and to reflect on future scenarios for public and private interaction.

The emergence of private forest carbon markets

Beginning in the early 1990s, businesses and civil society organizations created carbon offset projects focused on reducing emissions from forests, representing the first ever carbon offsets and instigating the birth of the voluntary market (Hamilton, Chokkalingam and Bendana 2010b). Though the interest in forestry carbon offset projects waned from the late 1990s until the early 2000s, global attention by the latest 2000s has redoubled. Along with the resurgence of interest, a multitude of voluntary certification programs has emerged that attempt to govern carbon projects through the setting of performance standards and the establishment of systems for third-party monitoring and verification of emission reductions. These include schemes focused exclusively on carbon dioxide mitigation as well as others that address "co-benefits", i.e. non-carbon environmental and social objectives, such as the protection of biodiversity or strengthening of community livelihoods. In Table 4.1 we highlight six of the major carbon offset certification standards now in operation, plus two major co-benefit standards.

There are several reasons for the renewed focus on forests. First, it is widely hypothesized that forest mitigation efforts may be more cost effective than efforts to reduce fossil fuel emissions (e.g. Eliasch 2008; Olsen and Bishop 2009; Stern 2006). Second, and partly as a result, global investments in reducing emissions from forests may continue to grow despite the possible failure of many industrialized countries to commit to steep non-forest emissions reduction targets in a post-2012 climate agreement. Forests are thus viewed in part as a way to "buy time" before the developed world can agree on more costly and drastic reductions associated with their own domestic emissions. While initially the inclusion of forests under the UNFCCC was restricted to afforestation and reforestation projects under the Clean Development Mechanism (CDM), the recent trend has been to further expand the integration of forests and other related activities, such as conservation of forests, into the international climate regime. Given that emissions from deforestation and forest degradation are roughly 12 per cent of global emissions (van der Werf et al. 2009), advocates argue that such emissions must be addressed irrespective of other strategies adopted. As a result, reduced emissions from deforestation and forest degradation in developing countries (REDD) and forest enhancement (REDD+) has gained increased traction as a means to incentivize related activities. Third, the linkage of forest conservation

Table 4.1 Carbon offset verification standards.

Carbon offset standards		
Standard	**Project type eligibility**	**Project location eligibility**
American Carbon Registry Forest Project Standard	Afforestation and reforestation, integrated forest management, avoided deforestation	United States or non-Annex I countries
CarbonFix Standard	Afforestation and reforestation	No restrictions
Climate Action Reserve	Reforestation, avoided forest conversion, conservation-based forest management, urban forestry	United States (with planned extension to Mexico and Canada)
ISO 14064	No restrictions	No restrictions
Plan Vivo Standards	Afforestation and reforestation, agroforestry, avoided deforestation, forest conservation and restoration	Community-based projects in developing countries
Voluntary Carbon Standard	Afforestation, reforestation and revegetation, agricultural land management, improved forest management, avoided deforestation	No restrictions

Social or environmental co-benefit standards		
Standard	**Project type eligibility**	**Project location eligibility**
Climate, Community and Biodiversity Standards	Reforestation and afforestation, revegetation, forest restoration, agroforestry and sustainable agriculture, avoided deforestation	No restrictions
Social Carbon	No restrictions (as long as accredited methodology for that project type is available)	No restrictions

Sources: Hamilton, K., U. Chokkalingam, and M. Bendana (2010b) *State of the Forest Carbon Markets 2009: Taking Root and Branching Out*. Washington, DC: Ecosystem Marketplace; ⟨http://www.carbonfix.info/⟩; ⟨http://www.climateactionreserve.org/⟩; ⟨http://www.planvivo.org⟩; ⟨http://www.v-c-s.org⟩; ⟨http://www.climate-standards.org/⟩; ⟨http://www.socialcarbon.org/⟩; ⟨http://www.climate-standards.org⟩ (accessed April 23, 2010)

with global climate initiatives is seen by many forest stakeholders as a last chance to create incentives for forest protection, given the failure to develop a binding global forest regime. As we shall see in Chapter 5, this is a very different situation than, for instance, in the food sector, analyzed by Doris Fuchs and Frederike Boll, as the food sector has not achieved a "high-politics status" in relation to climate issues. Fourth, and most importantly, voluntary and private programs have been able to move much more swiftly to promote reductions in the forest sector compared to intergovernmental processes.

As a result, the voluntary arena is the primary location in which forestry offsets are currently being bought and sold. As of 2009, 73 per cent of all forestry offset transactions by value occurred in what is referred to by Hamilton et al. (2010a) as the voluntary carbon market. However, reflecting Karsten Ronit's conclusion to this volume about the importance of assessing degrees of "privateness", "voluntariness" and "programness", the difference between these "voluntary" programs and the intergovernmental programs that they refer to as the "regulated" forest carbon markets is much more nuanced. Hamilton et al. note that "voluntary" certainly captures all carbon offset trades that are not governed or mandated by a single sovereign state or intergovernmental agreements, such as the EU Emissions Trading Scheme (ETS).

However, governments can often act to affect or create the institutional conditions through which voluntary purchase of credits occurs, similar to the way the US Department of Agriculture determines what qualifies as an organic product, but leaves it up to consumers to decide whether or not to purchase these goods. In this regard governments are also involved in mandating procedures under which "over-the-counter" (OTC) offset markets conduct their business. Likewise governments play critically important roles in developing voluntary carbon exchange markets, such as the role they played in the Chicago Climate Exchange (CCX).[3]

While decisions to join CCX were purely voluntary, the nature of the exchange itself was subject to government oversight and scrutiny; participants, once they joined, were governed by mandatory contractual cap-and-trade emissions (Hamilton et al. 2010b). And in another twist, both governments and private actors were involved in providing "certified emission reductions" (CERs)[4] whose sole purpose was to verify whether programs actually provide the purported emissions reductions. For the purposes of this chapter, we distinguish purely public from private voluntary programs on the basis of whether *purchases* of carbon credits are governed by states (see Table 4.2).

Currently purchasers of forest carbon from voluntary markets range from individual consumers (the smallest segment), to private firms, to

Table 4.2 Volume and value of private and public forest carbon markets.

	Volume (Mt CO_2)		Value (US$ million)	
Markets	Historical total	2008	Historical total	2008
Voluntary markets[1]	17.9	5.0	137.6	36.8
Government mandated[2]	2.9	0.2	11.6	0.3
Total global markets	20.8	5.3	149.2	37.1

Source: Hamilton, Chokkalingam and Bendana (2010b)
1. Includes the "over-the-counter" market and the Chicago Climate Exchange
2. Includes New South Wales, CDM, New Zealand ETS and Kyoto markets, referred to by Hamilton et al. as "regulated"

non-profit, public and financial institutions. Transactions may involve project developers, wholesalers, retailers and/or brokers (Bayon, Hawn and Hamilton 2006). The global nature of voluntary carbon transactions is such that buyers from any location can purchase offset credits from projects all over the world. This diversity of actors and these transaction spaces create conflicting demands for innovative and diverse approaches as well as for the global standardization of procedures and performance requirements.

Despite the relatively rapid, and ongoing, growth and development of voluntary forest carbon markets, forestry credits represent a diminishing share of the overall market, with other sectors, such as energy efficiency and renewable energy projects, accelerating at faster rates.[5] Nevertheless, the overall transaction volume of forest carbon markets continued to rise until 2009, when the global financial crisis and uncertainties regarding compliance led to a fall in regulated and voluntary carbon trading across all sectors (Hamilton, Sjardin, Shapiro and Marcello 2009).

Estimating and accounting for emission reductions associated with forestry offset projects faces numerous challenges, including issues of "quality" relating to additionality, permanence and leakage. "Additionality" refers to the degree to which project activities result in the sequestration/avoided emissions more than would otherwise have been sequestered in the absence of such activities; "permanence" refers to the degree to which the project's carbon sequestration/avoided emissions will be maintained over the long term; and "leakage" refers to whether or not sequestering carbon in one project area will simply shift or "leak" deforestation/degradation activities to another location outside the project's bounds.

Following initial delays, a range of carbon offset standards organizations have developed methodologies aimed at ensuring that these carbon quality issues are addressed. These methodologies shape the behavior of the project hosts and developers, the businesses or organizations involved in the creation of the offset credits. Development of the methodologies served as a notable barrier to the uptake of forest offset projects in the voluntary sector, as their development differed substantially from renewable energy or energy efficiency projects, and so required significant innovation and substantiation. The American Carbon Registry Forest Project Standard, an offshoot from the United States's first private voluntary GHG emissions registry, made important strides in this regard by developing methodologies for A/R, IFM and REDD projects in 2007. In 2002, third-party monitoring and verification of forest carbon schemes accounted for only 15 per cent of all forest carbon offsets but this figure had reached a staggering 96 per cent as of 2009 (Hamilton et al. 2010b). In other words, third-party effort has now become a de facto prerequisite for voluntary carbon reduction markets, and the methodologies are available for application. Thanks to the norms developed in the voluntary market, the collective action dilemma for getting firms to join (pointed to in Chapter 1) thus appears to have been overcome in this case.

As Table 4.1 reviews, a number of private standards now verify forest carbon projects. Carbon offset standards such as VCS are designed primarily for the accounting of the GHG emissions reductions or enhanced removals associated with the project. Standards such as CCBA and Social Carbon, on the other hand, are designed to address a broader set of goals, primarily for additional social or environmental co-benefits generated by the projects. These co-benefits standards, therefore, can be doubled up with carbon offset standards to demonstrate that a single project delivers both sets of benefits.

Not only are states and intergovernmental arenas important for providing clarity and rules governing the emergence of markets that rely on voluntary decisions by various customers; intergovernmental and state-based efforts to reduce climate emissions are also critical for understanding the ways private and public programs intersect and co-evolve. To pave the way for illustration of such interaction, we first review relevant intergovernmental processes.

Intergovernmental processes relevant to forests and climate

Our review of intergovernmental processes contains three steps. These illustrate some of the important institutional conditions surrounding forestry and climate change programs. First, we review how intergovernmental efforts have developed parameters for developed country accounting of carbon emissions and sinks resulting from changes in forest cover. The following section illustrates how these efforts, referred to as land use, land use change and forestry (LULUCF) within the Kyoto

Protocol, have had profoundly important ideational impacts on both public and private programs.

As a second step we review the Kyoto Protocol's Joint Implementation (JI) and Clean Development Mechanism (CDM), market mechanisms that allow for carbon offset arrangements between developed and developed and developing country actors, respectively. The CDM mechanism has been important for providing greater understanding about what projects, public or private, might be considered as appropriate for emissions reduction, as well as for creating an environment that is synergistic with private programs.

Third, we then review the proliferation of efforts to include reduced emissions from deforestation and degradation and forest enhancement (REDD+) in developing countries within a post-2012 climate agreement. Such REDD+ efforts have had two key impacts on private programs. They have created a shift towards the use of public or private "funds" as a way to create incentives to reduce emissions, and they have led the championing of existing private certification programs as prerequisites to any REDD+ programs.

Credits for emissions and removals from LULUCF activities

The Kyoto Protocol has had a great influence in shaping ideas about what types of methodologies are appropriate for assessing emissions reductions/enhanced sequestration associated with the land use sector, which in turn has shaped ideational choices within private voluntary programs about what, precisely, constitutes an emission reduction. The key part of the Kyoto Protocol of relevance here are the requirements for Annex I Parties to account for carbon sources and sinks resulting from land use, land use change and forestry activities (LULUCF) carried out *within* each country (articles 3.3 and 3.4, Kyoto Protocol). Net changes in such emissions and removals resulting from afforestation (creation of forests on lands that have been out of forest use for at least 50 years), reforestation (re-establishment of forests on land previously forested that had been converted to non-forest use and lacked forest on December 31, 1989) and deforestation (direct, human-induced conversion of forested land to non-forested land) activities can be included in Parties' emission reduction commitments. Industrialized nations have the option of including forest management in inventory accounting, as pursuant to article 3.4 of the Protocol. Once a choice is made to include or exclude forest management, that choice is fixed for the first commitment period. Countries face a dilemma when deciding whether or not to include forest management, since natural disturbances, such as fire and pest outbreaks, are difficult to predict. This means that forest management could vary from a source to a sink of carbon dioxide emissions from year to year (Nelson 2003).

Credits from JI and CDM activities

The Kyoto Protocol has also had an important role in shaping two types of interventions aimed at reducing the costs of emissions reductions. These include "Joint Implementation" (JI) in which an Annex I country can offset some of its domestic emissions by investing in a project in another Annex I country or a country in economic transition, in exchange for emission reduction credits. These projects span sectors, but include projects in forest management, reforestation and afforestation. One key principle for JI projects is that they must be developed in accordance with the rules, modalities and guidelines under LULUCF to ensure additionality and mitigate against leakage.

The second type of cost reduction intervention created under the Kyoto Protocol involved the offsetting of Annex I, or developed countries', emissions with investment in projects in non-Annex I countries, or developing countries. The institutional arena through which such projects are approved is known as the Clean Development Mechanism (CDM). As we discuss below, the CDM is important for our analysis because of its synergistic potential with private "Gold Standard" certification, and also for its normative and ideational impact in developing notions about appropriate land use projects.

A key decision under the CDM initially was that LULUCF activities be limited to afforestation and reforestation to qualify for temporary offset credits. Reduction of emissions from deforestation and forest degradation did not qualify under the first commitment period because of concerns raised during the Marrakech negotiations. Official explanations focused on methodological concerns, such as those related to leakage, additionality, permanence and baselines (IPCC 2007). Additionally, a number of organizations were concerned that such efforts would take pressure off countries to reduce their industrial emissions (Boyd, Corbera and Estrada 2008). The combined result of restricting forest carbon projects to afforestation and reforestation, and the many limitations and requirements placed on approval of those A/R projects, meant that the market for forest carbon under the CDM has been very slow to develop (Cerbu, Swallow and Thompson 2011). This helps explain, as we discussed above, why the private forest carbon market emerged during this time as a relatively flexible and agile arena in which to accommodate forest carbon trade.

At the same time scientific research about the role of forests in emitting carbon dioxide, and the relatively low cost of reducing forest carbon emissions, led many public and private interests to change their initial opposition and support a more inclusive treatment for a second commitment period. This, in turn, reopened the negotiating process for the CDM to allow for the possibility of including a wider range of forestry projects, and shaped new political conditions for private initiatives. As a result, project-level activities that reduce deforestation and forest degradation are now being considered under the CDM and/or through new mechanisms expressly designed for reduced emissions from deforestation and forest degradation (REDD).

The emergence of REDD and REDD+

Despite many concerns over the mechanics of including natural forests in a global climate regime, increasingly large coalitions of environmental groups, industry organizations and government officials are now indicating their support for REDD as the latest international effort designed to "bring forests back in". REDD is focused explicitly on forest resources within non-Annex I Parties, which have the lion's share of tropical forest area and forest ecosystem biodiversity. Knowing why this has occurred is critically important for understanding, as we discuss next, the voluntary and private carbon market, as well as the intersecting role of private mechanisms with public REDD initiatives.

A key galvanizing event that pushed REDD onto the international agenda was the 2005 Conference of the Parties (COP 11) meeting in Montreal when Papua New Guinea and Costa Rica led several other nations, collectively known as the "Coalition for Rainforest Nations" (Myers 2007), in a formal request to place REDD on the climate agenda (UNFCCC 2005). This interest, spurred by scientific data and economic analyses, was viewed as a way to entice developed countries to commit the resources and technical assistance needed to compensate developing countries for revenues forgone by slowing and/or reversing the conversion of forests to other land uses, and to enable them to pursue more sustainable rural development options. Developed countries responded with strong interest, with COP 11 participants agreeing to have the standing scientific advisory committee, the Subsidiary Body of Scientific and Technological Advice (SBSTA), review the prospects for including forests under climate negotiations.

The political momentum to include REDD in a climate regime gained further traction at the December 2007 climate negotiations in Bali (Appleton et al. 2009). Deliberations were now focused not on whether, but on how, to include forests, and how to promote a broad array of forest-related activities, now captured under REDD+ proposals (Appleton et al. 2009). Significant progress was made in identifying the resources developing countries would need for capacity building including monitoring, reporting and verification. The UNFCCC produced a document that outlined a two-year roadmap to prepare for the 2009 Copenhagen meeting, and announced the creation of an Ad-hoc Working Group on Long-term Cooperative Action (AWG-LCA) under the Convention to consider, among other issues, "positive incentives for developing countries to participate and to assess the role of conservation, sustainable management of forests and enhancement of forest carbon stocks in developing countries" (UNFCCC 2008).

The AWG-LCA focused on four key questions, among others. First, it assessed whether REDD's goals should cover only deforestation and forest degradation or also conservation and sustainable management of forests (REDD+). This is important because it affects the degree to which intergovernmental climate institutions can comprehensively address forest conservation and thereby compensate for the limited direct impact of international forest institutions reviewed above. It also affects the role of voluntary mechanisms within these efforts.

Second, there have been vigorous discussions about whether efforts should be focused on project-level reductions, which some assert might more directly benefit indigenous and local communities, or focused on national policies and programs. Much of these debates are concerned with whether a national focus would create more logical baselines from which reduced emissions could be calculated, and help avoid double counting and leakage that can occur where sequestration in one project might inadvertently lead to increased emissions elsewhere. Still others argue that subnational delineations based on forest ecosystems are often better than national "lines on a map" for calculating baselines.[6]

Third, and reflecting neoliberal norms reinforced by the emergence of private and voluntary programs, parties to the UNFCCC have considered two key policy mechanisms: market- and fund-based mechanisms. Deliberations focused on the pros and cons of incorporating REDD activities into a cap-and-trade scheme versus a fund established to directly finance forest emission reduction efforts. While on the one hand a market-linked mechanism may provide a more adequate, stable flow of finance, a fund-based instrument can protect sovereignty, ensure that emissions are reduced in non-forest sectors, and be used for capacity-building efforts that do not result in immediately quantifiable emissions reductions. However, there are also proposals to make the REDD mechanism "market linked" in which a steady stream of finance would be generated by the carbon market but credits generated from REDD would not be fully fungible with non-REDD credits.

Fourth, Parties deliberated over how to precisely monitor emissions reductions, also an important issue in private voluntary programs. These deliberations focused on whether spatially explicit data would be required; specific methodologies for interpreting this data; the monitoring of non-carbon attributes; and the degree to which safeguards should be established for biodiversity and indigenous and other forest-dependent communities.

The 2009 Copenhagen Accord, affirmed by the 2010 Cancun Agreement, verified global support for a "REDD+" mechanism. The decision was also made to establish a Green Climate Fund designed to support REDD+ activities (Appleton et al. 2009; UNFCCC 2011). As part of these efforts, developing countries were formally requested to identify drivers of deforestation, establish national forest management systems, develop guidance for engagement of indigenous peoples and local communities in monitoring and reporting, and develop forest reference emission levels that take into account historical data and adjust for national circumstances. This approach also raised the idea that funds, be they public or private, might act as a way to achieve reductions through particular projects in potentially efficient and effective fashion.

The way in which REDD will be treated in a post-2012 climate has also raised important questions on how the interaction of public and private authority might help address the thorny trade-offs among desired goals such as national sovereignty and rights to economic development and diversification, biodiversity conservation, food security, and the provision of local subsistence and livelihood needs. Forests in many developing countries are integral to the establishment and defense of land tenure and resource rights, and thus play a pivotal role in the national political economy. The most heavily forested areas are often occupied by traditional and indigenous societies, while the forest edges may contain large populations of displaced and marginalized subsistence farmers. Forest tenure in such contexts is often unresolved or insecure, contributing to low land prices.

For these reasons, some contenders fear that the monetization of forest carbon, and particularly natural forest carbon, will spur a "great land grab", displacing the rural poor that depend upon these lands as their source of livelihoods, and favoring "simplified" tree plantations over the maintenance of forest biodiversity. Takacs (2009) argues this has already happened under the CDM, where resource and/or land rights were granted to private investors in China and Tanzania to establish tree plantations on degraded lands, thereby excluding local land users. The projects claimed to meet co-benefit objectives by hiring local people as laborers and restoring damaged landscapes. However, plantation labor is generally seasonal, pay is quite low, and degraded lands may provide more important native habitats than the tree plantations that replace them.

It is for these reasons that REDD+ proposals have emerged, but they, too, have attracted critiques. Some fear the corrupt use of REDD+ funds owing to: the low governance capacity in REDD eligible tropical countries; inadequate recognition of indigenous peoples' rights; failure to effectively engage indigenous and local communities in REDD+ design; perverse incentives to convert natural forests to tree plantations in order to facilitate carbon accounting and measurement; perverse incentives to increase deforestation rates in the short term in order to demonstrate greater "reductions" in the long term; technical measurement and monitoring challenges; corruption; flooding of the market with cheap carbon credits; as well as other concerns common to all carbon offset trading, such as reduced incentives for developed countries to lower their fossil fuel emissions (REDD-Monitor 2010).

The inclusion of sustainable forest management projects under the REDD+ umbrella is another source of substantial controversy. Much of the critique is focused on industrial timber harvest, particularly within primary natural forest. The core of this avenue of critique is that logging of natural tropical forests represents a net loss of carbon, biodiversity and local socioeconomic welfare and that incentives to "reduce" these impacts will instead serve to validate and promote unsustainable exploitation (Global Witness 2009).

As a result many have turned to private authority, especially certification schemes, as means to address indigenous rights and ecosystem management. As discussed below, several such schemes explicitly focus on the environmental and social dimensions of land management, including the Climate, Community and Biodiversity Alliance (CCBA) and Plan Vivo. To some stakeholders, certification under such schemes should be established as a prerequisite before any project could be considered for REDD+ funding.

The potential for direct, symbiotic and learning impacts from public and private programs

What is the ability of private forestry programs to contribute to climate mitigation, while safeguarding environmental and social goals related to forest governance? Our analytical framework points to a series of pathways through which the contribution could extend beyond the direct effects of immediate emissions reductions. It is these other pathways, where private and public realms intersect in novel ways, that suggest a more promising outlook than an assessment of direct effects alone, and point to different potentials.

Direct impacts from private voluntary programs

The direct effects of voluntary forestry programs on the problem of global climate change indicate that the impact is minute. In 2008, global markets transacted a volume of 4840 Mt CO_2e (CO_2 equivalent), of which 5.0 Mt CO_2e represents the voluntary forest carbon trade (Hamilton et al. 2010a, 2010b). Voluntary forest carbon offsets, in other words, represent a fraction of a per cent additional contribution to the global public climate mitigation efforts, at least in terms of direct impact.

Nevertheless, if we consider the relative direct impacts of private forest carbon projects in comparison to intergovernmental forest carbon initiatives, the differences are stark. Initially, private carbon forestry projects experienced dramatically more rapid growth, distribution and diversity of project types as compared to their counterparts within the CDM or JI mechanisms. The first CDM forestry project, located in China, was approved in 2005 (Carbon Positive 2005), followed by a delay of four years before the approval of the second project, located in India (Carbon Positive 2009). (However, the approval rate accelerated in 2009, with a total of 14 projects approved by April 2010; McNeil 2010.) This suggests that private markets have thrived in spite of, or perhaps because of, the controversies that prevent the full inclusion of forests within the intergovernmental climate arena.

As reflected in Table 4.2, the voluntary carbon market has changed rapidly in the last few years as it has grown and matured, as has the role of forestry within the market. In 2008, 5603 kt CO_2e of land-based credits were transacted in the voluntary OTC market (Hamilton et al. 2009). As of 2009, forest carbon finance was estimated to affect 2.1 million hectares of land and was valued at approximately US$149.2 million (Hamilton et al. 2010b). The projects generating these revenues vary considerably in scope, scale and type of forest management activity, as well as in the range of actors involved. Understanding this diversity is essential for understanding the social dynamics and controversies that forest carbon trading entails, as well as to assess its net impacts across the landscape and over time.

Meanwhile, however, intergovernmental rule-making under the CDM and JI, and increasingly in anticipation of REDD, have significantly influenced the design processes of both domestic forest policy and private voluntary schemes (Angelsen, Brockhaus, Kanninen and Sills 2009; Boyd, May, Chang and Veiga 2007; Corbera and Brown 2008). The shadow of this intergovernmental rule-making has positively fertilized the ground for private voluntary forest carbon schemes. As of 2009, the voluntary markets for forest carbon had generated over 200 projects in 40 different countries (Hamilton et al. 2010b). From the perspective simply of distribution and reach, the obstruction of forest carbon programs within the intergovernmental arena thus appears to have heightened the relative importance of private voluntary programs. Some have suggested, furthermore, that the potential for voluntary markets to avoid the burdensome administrative requirements of CDM forestry projects may also facilitate access for smaller-scale producers (Peskett, Luttrell and Iwata 2007).

However, the lack of a single common registry and the diversity of relevant performance standards and requirements make it relatively difficult to assess the net environmental and social impacts of carbon forestry within the voluntary arena. In other words, there is much activity but still limited coordination, one of the typical pitfalls of private voluntary programs. A recent comparison (Corbera, Estrada and Brown 2009) of CDM and voluntary carbon offset schemes suggests that projects within the CDM covered a broader geographic range and included more small-scale operations (presumably facilitating local benefit) than those under voluntary schemes. This study also found greater variation in design and in the environmental and social standards applied to purely voluntary schemes (Corbera et al. 2009). These findings suggest that, while private schemes may allow more room for policy innovation, many may be constrained by market forces favoring economies of scale, and lack guarantees that carbon trading meets a set of minimum standards. These researchers, therefore, questioned the ability of private markets alone to guarantee sufficient and/or desirable environmental and social performance.

Public–private symbiotic pathway

In this section, we discuss the emergence of public–private symbiosis. This can be created when there are governance gaps, whereby the public authority does not adequately address various goals and private schemes are layered on top to address such gaps. We use the example of the CDM Gold Standard, analyzed from a different perspective in Chapter 2 by Jennifer Clapp and Jason Thistlethwaite, to illustrate this possibility. While the CDM Gold Standard currently excludes forestry projects, if expanded or if other programs were to develop similar symbiotic relationships with public authorities, this could enhance the strength of such pathways.

The CDM Gold Standard was developed in response to the perceived weaknesses of the CDM in addressing non-carbon attributes of climate change offset projects. According to the Kyoto Protocol language, an additional purpose of the CDM – beyond assisting Annex I Parties in meeting their reduction targets – is to promote sustainable development benefits in non-Annex I Parties (UNFCCC 1997). Article 12 of the Kyoto Protocol states that:

> the purpose of the Clean Development Mechanism shall be to assist Parties not included in the Annex I in *achieving sustainable development* [authors' emphasis] and in contributing to the ultimate objective of the Convention, and to assist Parties included in Annex I in achieving compliance with their quantified emission limitation and reduction commitments under Article 3. (UNFCCC 1997)

The impetus behind inclusion of sustainable development benefits was the realization that projects invested in developing countries have the potential to not only reduce GHG cost effectively, but also contribute to the improvement of conditions in the developing nations (Boyd, Hultman and Timmons Roberts 2009).

In practice, however, the combined challenge of achieving both measurable emissions reductions and sustainable development, and in a manner credible to a wide diversity of global stakeholders, is formidable indeed. Since its inception, the CDM program has generated considerable criticism, varying from complaints in regard to the lengthy credit issuance process to the lack of standardized baseline and monitoring methodologies to the deficiency of financial and personnel support for the CDM executive board. Partly in connection with these technical difficulties, many have criticized the CDM for its failure to generate sustainable development benefits (Cosbey et al. 2005; Kenber 2002).

With recognition of CDM's perceived shortfalls in achieving its sustainable development goal, a group of civil society organizations launched the non-profit organization, the Gold Standard, in 2003. In contrast to purely private offset programs, the Gold Standard is a private certification scheme designed with the express intention of strengthening the non-carbon attributes, including the treatment of sustainable development benefits, of the intergovernmental CDM. That is, the Gold Standard certification is layered on top of a project approved by the CDM, strengthening rather than competing with the CDM structure. Thus, unlike traditional certification programs, the Gold Standard augments an existing public policy, but depends upon this governmental structure for its functioning. That is, if the CDM was abolished, the status of the Gold Standard as a mechanism to internalize CDM externalities would be ambiguous at best. The relationships between the public and private markets are increasingly interwoven, so that – like the Gold Standard and CDM – their existence may rely to a greater degree on one another, and thus shape conditions for compliance.

Importantly, the Gold Standard certifies only renewable energy and energy efficiency projects, to the exclusion of forestry and land use projects. However, one could envision a Gold Standard-like public–private symbiosis involving certification schemes and a future REDD intergovernmental program.

Private–private symbiotic pathway

While the Gold Standard does not yet address forestry projects and thus has had little direct impact within the forest sector, we point to a second symbiosis pathway with specific application to forest carbon voluntary programs. As discussed above, among the various standards operating in the voluntary forestry carbon arena, there is a division between those standards designed to account for a project's emissions reductions, and then there are a smaller set of standards designed to test for additional social or environmental co-benefits generated by the projects.

Since these co-benefit standards are stacked on top of the carbon offset standards, to indicate to the consumer that a given project provides both kinds of benefit, we describe this phenomenon as private–private symbiosis. Many of the broad-based schemes prioritize credible forest carbon verification and give relatively cursory attention to environmental and social impacts. For example, co-benefit guidelines in the VCS are limited to generalized instructions to "identify" and "undertake steps to mitigate" environmental and social impacts, and no social or environmental monitoring is required (Merger and Williams 2008).

In contrast, a number of schemes which focus exclusively on the land use sector place greater emphasis on co-benefits. CarbonFix, a German-based organization established in 1999 by scientists and other technical experts, is designed expressly for forestry projects only (specifically, afforestation and reforestation). The CarbonFix standard emphasizes carbon accounting and the creation of tradable credits, and also includes a rating system for ecological and social co-benefits. The scheme is now recognized as meeting the criteria for carbon credits under the German Emissions Trading Authority (CarbonFix 2010).

The Climate, Community and Biodiversity Alliance (CCBA), founded in 2003, is based in the US and consists of civil society groups, advising research institutes and corporate sponsors. The stated goal of CCBA is to provide "integrated solutions to land management" that mitigate climate change, support sustainable development and conserve biodiversity. CCB standards outline detailed environmental and social requirements, and systems for their monitoring and verification. Various levels of achievement are recognized through the awarding of standard, silver or gold certificates (CCBA 2010).

Plan Vivo, another set of standards, emerged in 1994 from a research project in Chiapas, Mexico. Plan Vivo is based in the UK and directed by a board of civil society representatives, scientists and technical consultants. The stated goal of Plan Vivo is to provide a framework for community-based land use projects that generate long-term carbon, livelihood and ecosystem benefits to smallholders and forest-dependent communities

(Plan Vivo 2010). Plan Vivo places greater emphasis on community participation and involvement than on formal, standardized verification procedures (Merger and Williams 2008).

The relationship between the carbon offset standards organizations and those verifying co-benefits appears to be synergistic. Many forestry-specific standards expressly refer to other schemes that can provide complementary coverage, thereby acknowledging their different foci and strengths. For example, CCBA encourages operations to also be certified under VCS as a means to more rigorously measure carbon leakage, permanence and additionality. VCS, in turn, recognizes CCBA as providing a framework for assessing co-benefits. Such synergies have both negative and positive implications, as the diversity of schemes and standards may cause confusion.

On the one hand, demand for multiple certifications contributes to high transaction costs where much of the financial benefits may be captured by intermediaries. On the other hand, multiple schemes provide multiple venues for stakeholder participation and thus may address the needs and concerns of a wider diversity of stakeholders, while providing market differentiation for varying levels of environmental and social performance. In fact, Hamilton and colleagues find that in the first two quarters of 2009, 35 per cent of forestry project credits were validated by a stacking of VCS, CCBA and the American Carbon Registry (Hamilton et al. 2010b).

The potential for learning

Finally, in regard to our third impact pathway, i.e. policy learning, the role of private voluntary programs is virtually indisputable. Voluntary programs for carbon offsets have provided a multitude of sites for innovation on technologies and accounting methodologies, as well as approaches to addressing highly controversial environmental and social forestry issues. Kollmuss, Zink and Polycarp (2008: 13) for instance explain: "A key role of the voluntary market is to shape the rules and procedures for offsets in future compliance markets. In other words, the voluntary market can be used as a testing ground for procedures, methodologies and technologies." Thus, these learning processes can inform the different elements of designing, joining, monitoring and complying.

Particularly in the United States where no federal climate change regulatory market yet exists, the market for voluntary carbon offsets plays a critical role in testing alternative designs and in coalescing participants around the various options (VCS brochure). Standards, registries and protocols are developed explicitly to serve as pre-compliance standards. This is especially the case for the "commodity standards". For example, the Climate Action Reserve's forestry protocols, which specify the procedure through which to quantify the sequestered carbon from a forestry project, are viewed by some as frontrunner models for forest offsets in California and/or federal climate legislation (Rose 2008).

Further, an iterative relationship between public and private carbon mitigation programs is becoming evident, so that not only are elements from the voluntary market introduced to regulatory schemes, but government-led programs have influenced the institutional specifics of voluntary programs. The design of California's cap-and-trade program has relied on innovative methodologies from the voluntary market. On the other hand, a recently developed mosaic REDD methodology by Terra Global Capital is based partly on CDM methodologies for afforestation/reforestation. This approach is notable for addressing "mosaic" deforestation in which multiple drivers account for the occurrence of deforestation (Peters-Stanley 2010).

Projecting forward

The complex relationship between a private voluntary program and public authority suggests several possible "futures" for forest-related private voluntary climate programs. They may remain in a niche market, catering to government lead-by-example programs and other early actors. Or they could be formally integrated into the governmental program (i.e. the Gold Standard successfully transforms the CDM, and/or the CCBA transforms REDD+) with requirements of social and environmental benefits joining existing CDM project criteria (or national, subnational or project-level payments under REDD+). Under these scenarios, the private programs' governing powers would work formally in tandem with CDM and/or REDD+ criteria, or they could even, at some point, decide to divest themselves of private standard-setting bodies and become fully absorbed by the public authority of the CDM executive board or future REDD+ international, national and/or subnational governance regimes.

We focus our attention on another possible outcome, the maturation of the symbiotic relationship between public policy and the private voluntary program into a stable, complementary, reinforcing companion to governmental programs, and thus in a different way marry these programs with public goals. In this scenario, the regulatory market will come to treat environmental and social certification of credits as indispensable, just as environmental and social certification schemes will rely on compliance with regulatory market requirements. Most significantly, this would mean that parties seeking to promote policy objectives may do so by attempting to influence the private standard rather than the public policy process. Interestingly, with the assumption that credits generated under regulatory markets are more robust than voluntary credits, this could result in advances in climate policy within the voluntary sector. As a result, the regulatory markets would become tightened as credits bought on the voluntary market are retired to ensure the integrity of reductions.

This symbiotic relationship between public and private authority may be particularly critical for addressing deforestation and forest degradation, including REDD+ efforts under the UNFCCC, post-Kyoto. The expansion of carbon markets or compensation mechanisms into forested areas has heightened concerns surrounding unintended negative consequences for local communities and biodiversity. Particularly because addressing specific implementation practices may infringe on national sovereignty at the global negotiations level, the layering of a private voluntary program onto public policy may prove a more pragmatic avenue.

The learning pathway suggests ways in which the private and public markets may be linked through the transmission of innovations and knowledge. This points to an indirect effect of private programs that may improve climate problem-solving in the public realm, for example by developing new technologies and innovations that reduce costs and uncertainty of mitigation activities and so make multi-scale cooperation feasible and sustained (Hepburn and Stern 2008).

Conclusion

Our review identifies several important implications for understanding the impact of private authority on global forest and climate governance. First, it is critical that the actual problems for which interventions are developed are well understood so that policy analysts can move from assessing support to assessing environmental effectiveness. Second, it is key to apply a historical analysis that places in context why a private voluntary program emerged, and potential support for mechanisms that bypass public authority. Third, we highlight the importance of more carefully assessing the evolutionary potential and intersecting effects of public with private authority.

The role of forestry in voluntary climate programs, which may also pave the way for governmental interventions at a later date (Auld, Balboa, Bernstein and Cashore 2009), is a critically important yet poorly understood arena for learning. In addition, the symbiotic relationship between the private voluntary program with the public policy process offers several potential advantages and emphasizes different facilitators. This public–private relationship could be more efficient than state-based processes in achieving some of the voluntary program's objectives (Cloghesy

2004; Kirton 2004), as advancement of a voluntary program will not have to contend with hurdles of state sovereignty nor face other barriers associated with the policy process, such as the lack of resources and capacity for implementation and enforcement. Perhaps most significantly, the use of a voluntary private program in tandem with the public policy process bypasses hurdles inherent in reopening the negotiating process and possibly reengaging in hard-won battles. Thus, this model of shared public–private management of an environmental problem could hold lessons for the design of global environmental governance.

As Gunningham et al. argue, non-state involvement, in conjunction with government intervention, is often more effective than a single-instrument approach in achieving desired outcomes. They argue that a greater range of actors in policymaking can relieve governments' limited financial and personnel resources, while yielding broader support and legitimacy from civil society and those being regulated. They suggest that "together, and in conjunction with state action, they achieved far more than state action alone was ever likely to" (Gunningham, Grabosky and Sinclair 1998: 205).

What is certain is that any scholarly analysis of private authority, carbon and forests must be sure to develop an analytical framework sensitive to direct and indirect effects in this particular sector, the evolutionary potential of different instruments, and what appear to be the most promising "intersecting" effects. Identifying and harnessing these intersections, specifically in the form of symbiosis, may enable policymakers to contribute more effectively to reversing deforestation and forest degradation while mitigating global climate change.

Notes

1. Indonesia alone is currently contributing as much as 12 per cent of global CO_2 emissions (van der Werf et al. 2009). It is important to note, however, that the share of climate emissions associated with degradation and deforestation will ultimately become smaller compared to industrial pollution as the carbon is released. Hence, carbon is now viewed by many as important for "buying time" because emissions reductions are deemed less expensive than in the automobile sector.
2. Our thanks to Shelby Semmes for elaborating this point.
3. The fragile nature of these voluntary markets, which we explore in the conclusion, was recently illustrated with the decision to disband the Chicago Climate Exchange (*New York Times* November 18, 2010).
4. Unless specifically noted, our analysis of the "public" and "private" forest carbon markets contrasts the OTC and CCX markets with the intergovernmental CDM and emerging REDD+ mechanisms under the UNFCCC. We do not focus on other public emissions trading schemes involving forest credits, e.g. those of New South Wales or New Zealand. In sum, it is beyond the scope of this chapter to detail the complex linkages between all

regulated and voluntary carbon markets. Our intention, rather, is to highlight the interdependent and interactive nature of the wide diversity of these public and private institutions.

5. The market share of land-based credits was 11 per cent in 2008 compared to 16 per cent in 2007.
6. A key issue for national-level accounting concerns the crediting scheme, in which countries would calculate a baseline of historical deforestation rates and would assess the change in deforestation levels over time. The country could then account for generated emission reductions achieved when emissions fall below a predetermined reference level. While there are many proposals that fall under this category, one of the first was introduced by Papua New Guinea on behalf of the Coalition for Rainforest Nations. Under this proposal, entitled "Compensated Reduction", nations would elect to participate and then create a baseline for their rate of deforestation, much like they do for other emission reductions sectors, and reduce levels of deforestation in exchange for credits that would be fungible on the global carbon market (Myers 2007). Under another early proposal from the Center for Clean Air Policy's (CCAP) Dual Markets Approach, a predetermined percentage of national emissions targets would be met through REDD projects, programs and policies, while the other percentage would be generated from efforts to reduce emissions in other sectors. The design to limit the percentage of REDD credits aims to address methodological concerns, as well as concerns that REDD efforts could detract from energy system and behavioral transformation. Unlike Compensated Reduction, the REDD market would be kept separate, in an effort to avoid potential disruption and volatility (Ogonowski et al. 2007). A "nested approach" has been championed by the Centro Agronómico Tropical de Investigación y Enseñanza (Pedroni, Streck, Estrada and Dutschke 2007) which provides for a national baseline to be established, and credits can be generated if emission levels are reduced below a given reference year. At the same time, subnational project-level developers can hold credits, regardless of national performance on REDD, a design feature aimed to attract private investment. Projects can be developed irrespective of the country's participation in the scheme; however, above a threshold of project development, the country would be committed to a national target.

REFERENCES

Angelsen, A., M. Brockhaus, M. Kanninen, and E. Sills. 2009. *Realising REDD+: National Strategy and Policy Options*. Available at ⟨http://www.cifor.cgiar.org⟩.

Appleton, A., P. Doran, M. Gutiérrez, K. Kulovesi, M. Muñoz, and C. Spence. 2009. "Summary of the Thirteenth Conference of the Parties to the UN Framework Convention on Climate Change and Third Meeting of Parties to the Kyoto Protocol." *Earth Negotiations Bulletin* 12 (354): 1–2.

Auld, G., C. Balboa, S. Bernstein, and B. Cashore. 2009. "The Emergence of Non-State Market Driven (NSMD) Governance Systems Across Sectors." In *Governance for the Environment: New Perspectives*, eds. M. Delmas and O. Young, 183–219. Cambridge: Cambridge University Press.

Bayon, R., A. Hawn, and K. Hamilton, eds. 2006. *Voluntary Carbon Markets: A Business Guide to What They Are and How They Work.* London: Earthscan.

Boyd, E., E. Corbera, and M. Estrada. 2008. "UNFCCC Negotiations (pre-Kyoto to COP-9): What the Process says about the Politics of CDM-sinks." *International Environmental Agreements: Politics, Law and Economics* 8 (2): 95–112.

Boyd, E., N. Hultman, and J. Timmons Roberts. 2009. "Reforming the CDM for Sustainable Development: Lessons Learned and Policy Futures." *Science & Policy* 12 (7): 820–831.

Boyd, E., P. May, M. Chang, and F. Veiga. 2007. "Exploring Socioeconomic Impacts of Forest Based Mitigation Projects: Lessons from Brazil and Bolivia." *Environmental Science & Policy* 10 (5): 419–433.

Carbon Positive. 2005. *First CDM Forestry Project Approved* [online news bulletin]. Carbon News and Info, November 28, 2005. Available from ⟨http://www.carbonpositive.net/⟩.

Carbon Positive. 2009. *Second CDM Forestry Project Approved* [online news bulletin]. Carbon News and Info, March 9, 2009. Available from ⟨http://www.carbonpositive.net/viewarticle.aspx⟩.

CarbonFix. *CarbonFix Standard: Quality CO_2 Certificates from New Forests* 2010. Available from ⟨http://www.carbonfix.info/⟩.

CCBA. *The Climate, Community & Biodiversity Alliance,* [web page] 2010. Available from ⟨http://www.climate-standards.org/⟩.

Cerbu, G. A., B. M. Swallow, and D. Y. Thompson. 2011. "Locating REDD: A Global Survey and Analysis of REDD Readiness and Demonstration Activities." *Environmental Science & Policy* online preview: 1–13.

Cloghesy, M. 2004. A corporate perspective on globalisation, sustainable development, and soft law. In *Hard Choices Soft Law, Voluntary Standards in Global Trade, Environment, and Social Governance,* eds. J. J. Kirton and M. J. Trebilcock, 323–328. Aldershot, UK: Ashgate.

Corbera, E., and K. Brown. 2008. "Building Institutions to Trade Ecosystem Services: Marketing Forest Carbon in Mexico." *World Development* 36 (10): 1956–1979.

Corbera, E., M. Estrada, and K. Brown. 2009. "How do Regulated and Voluntary Carbon-Offset Schemes Compare?" *Journal of Integrative Environmental Sciences* 6 (1): 25–50.

Cosbey, A., J. E. Parry, J. Browne, Y. Dinesh Babu, P. Bhandari, J. Drexhage, and D. Murphy. 2005. *Realizing the Development Dividend: Making the CDM Work for Developing Countries* (Phase I Report), 1–84.

DeFries, R., F. Achard, S. Brown, M. Herold, D. Murdiyarso, B. Schlamadinger, and C. de Souza. 2007. "Earth Observations for Estimating Greenhouse Gas Emissions from Deforestation in Developing Countries." *Environmental Science & Policy* 10 (4): 385–394.

Eliasch, J. 2008. *Eliasch Review. Climate Change: Financing Global Forests*. London: UK Office of Climate Change.

FAO. 2006. *Global Forest Resource Assesment 2005: Progress towards sustainable forest managment*. FAO Forestry Paper 147.

FAO. 2007. *Country Information* 2007 [cited July 2007]. Available from ⟨http://www.fao.org/forestry/site/countryinfo/en/⟩.

Gaston, G., S. Brown, M. Lorenzini, and K. D. Singh. 1998. "State and Change in Carbon Pools in the Forests of Tropical Africa." *Global Change Biology* 4 (1): 97–114.

Geist, Helmut J., and Eric F. Lambin. 2002. "Proximate Causes and Underlying Driving Forces of Tropical Deforestation." *BioScience* 52 (2): 143–150.

Global Witness. 2009. *Vested Interests: Industrial logging and carbon in tropical forests*. Global Witness.

Gunningham, N., P. N. Grabosky, and D. Sinclair. Eds. 1998. *Smart Regulation: Designing Environmental Policy.* Oxford and New York: Clarendon Press and Oxford University Press.

Hamilton, K., M. Sjardin, A. Shapiro, and T. Marcello. 2009. *Fortifying the Foundation: State of the Voluntary Carbon Markets 2009*. Ecosystem Marketplace and New Carbon Finance.

Hamilton, K., M. Sjardin, M. Peters-Stanley, and T. Marcello. 2010a. *Building Bridges: State of the Voluntary Carbon Markets 2010*. New York and Washington DC: Bloomberg New Energy Finance and Ecosystem Marketplace.

Hamilton, K., U. Chokkalingam, and M. Bendana. 2010b. *State of the Forest Carbon Markets 2009: Taking root and branching out*. Ecosystem Marketplace.

Hepburn, C. and N. Stern. 2008. "A New Global Deal on Climate Change." *Oxford Review of Economic Policy* 24(2): 259–279.

IPCC. 2007. *Climate Change 2007: Synthesis Report,* p. 104. Geneva, Switzerland: World Meteorological Association.

Kenber, M. 2002. *Gold Standard: Quality Standards for CDM and JI Projects*. Basel: World Wildlife Fund.

Kirton, J. J., and Michael J. Trebilcock. Eds. 2004. *Hard Choices, Soft Law: Voluntary Standards in Global Trade, Environment and Social Governance*. Aldershot, UK: Ashgate.

Kollmuss, A., H. Zink, and C. Polycarp. 2008. *Making Sense of the Voluntary Carbon Market: A Comparison of Carbon Offset Standards*. Stockholm Environment Institute and Tricorona. Available from ⟨http://wwf.panda.org/about_our_earth/all_publications/?126700/A-Comparison-of-Carbon-Offset-Standards-Making-Sense-of-the-Voluntary-Carbon-Market⟩.

Levin, K., B. Cashore, and J. Koppell. 2009. "Can Non-State Certification Systems Bolster State-Centered Efforts to Promote Sustainable Development through the Clean Development Mechanism (CDM)?" *Wake Forest Law Review* 44: 777–798.

Malhi, Y., J. Timmons Roberts, R. A. Betts, T. J. Killeen, W. Li, and C. A. Nobre. 2008. "Climate Change, Deforestation, and the Fate of the Amazon." *Science* 319 (5860): 169–172.

McNeil, R. 2010. *New Forest Carbon Project is the "Swiss Army Knife" of Conservation* [online press release]. Conservation International, April 16, 2010. Available from ⟨http://www.conservation.org/newsroom/pressreleases/Pages/ColombiaCDMproject.aspx⟩.

Merger, E., and A. Williams. 2008. *Comparison of Carbon Offset Standards for Climate Forestation Projects Participating in the Voluntary Market: A Comparison of Climate, Community & Biodiversity Standard (CCBS), CarbonFix Standard (CFS), Plan Vivo Systems and Standard, and AFOLU Voluntary Carbon Standard (VCS)*. Christchurch, New Zealand: University of Canterbury.

Myers, E. C. 2007. *Policies to Reduce Emissions from Deforestation and Degradation (REDD) in Tropical Forests: An Examination of the Issues Facing the Incorporation of REDD into Market-Based Climate Policies*. Washington DC: Resources for the Future.

Nabuurs, G. J., O. Masera, K. Andrasko, P. Benitez-Ponce, R. Boer, M. Dutschke, et al. 2007. "Forestry." In *Climate Change 2007: Mitigation. Contribution of Working Group III to the Fourth Assessment Report of the Intergovernmental Panel on Climate Change*, eds. B. Metz, O. R. Davidson, P. R. Bosch, R. Dave, and L. A. Meyer, 543–578. Cambridge: Cambridge University Press.

Nelson, H., and I. Vertinsky. 2003. *The Kyoto Protocol and Climate Change Mitigation Implications for Canada's Forest Industry*. Vancouver: Sustainable Forest Management Network.

Ogonowski, M., N. Helme, D. Movius, and J. Schmidt. 2007. "Reducing Emissions from Deforestation: The Dual Market Approach." Washington DC: Center for Clean Air Policy (CCAP). Available at ⟨http://www.ccap.org/international/final%20redd%20report.pdf⟩.

Olsen, N., and J. Bishop. 2009. *The Financial Costs of REDD: Evidence from Brazil and Indonesia*. IUCN.

Pedroni, L., C. Streck, M. Estrada, and M. Dutschke. 2007. *The "Nested Approach": A Flexible Mechanism to Reduce Emissions from Deforestation*. Turrialba: CATIE.

Peskett, L., C. Luttrell, and M. Iwata. 2007. *Can Standards for Voluntary Carbon Offsets Ensure Development Benefits*? Specialist Series: Forestry Briefing. London: Overseas Development Institute (ODI).

Peters-Stanley, M. 2010. *California sees REDD in AB32 Future*. *EkoEco*, July 31, 2010.

Plan Vivo. *Plan Vivo: Improving Livelihoods, Conserving and Restoring Ecosystems* [web page] 2010. Available from ⟨http://www.planvivo.org/⟩.

REDD-Monitor. *REDD: An Introduction* [web page] 2010. Available from ⟨http://www.redd-monitor.org/redd-an-introduction⟩.

Rose, T. 2008. Is the Future of Forest Offsets Already Here? *The Katoomba Group's Ecosystem Marketplace*, August 22, 2008.

Rudel, T., R. DeFries, G. P. Asner, and W. Laurance. 2009. "Changing Drivers of Deforestation and New Opportunities for Conservation." *Conservation Biology* 23 (6): 1396–1405.

Santilli, M., P. Moutinho, S. Schwartzman, D. Nepstad, L. Curran, and C. Nobre. 2005. "Tropical Deforestation and the Kyoto Protocol." *Climatic Change* 71: 267–276.

Stern, N. 2006. *Stern Review on the Economics of Climate Change: Executive Summary*. London: UK Department of Energy and Climate Change.

Takacs, D. 2009. "Carbon into Gold: Forest Carbon Offsets, Climate Change Adaptation, and International Law." *West-Northwest Journal of Environmental Law and Policy* 15 (1): 39–88.

UNFCCC. 1997. UN Doc FCCC/CP/1997/7/Add.1, December 10, 1997; 37 ILM 22 (1998). United Nations Framework Convention on Climate Change.

UNFCCC. 2005. Reducing emissions from deforestation in developing countries: Approaches to stimulate action. Submissions from parties (Papua New Guinea and Costa Rica). Conference of the Parties, eleventh session. Montreal, 28 November to 9 December 2005.

UNFCCC. 2008. Decision 1.CP.13 para. 1b (3). United Nations Framework Convention on Climate Change.

UNFCCC. 2011. *The Cancun Agreements Dec 1/CP.16*. United Nations Framework Convention on Climate Change.

van der Werf, G. R., D. C. Morton, R. S. DeFries, J. G. J. Olivier, P. S. Kasibhatla, R. B. Jackson, B. J. Collatz, and J. T. Randerson. 2009. "CO_2 Emissions from Forest Loss." *Nature Geoscience* 2: 737–738.

VCS. Brochure. Voluntary Carbon Standard. Available online at: ⟨http://www.v-c-s.org/docs/VCS_brochure.pdf%3E⟩.

5

Emerging private voluntary programs and climate change: The blind spots of the agrifood sector

Doris Fuchs and Frederike Boll

Introduction

"Agriculture is one of the largest emitters of greenhouse gases" according to Greenpeace in its 2008 report, *Cool Farming*. Considering the direct (soil and livestock) and indirect (fossil fuel use in farm operations and the production of agrochemicals) emissions from the agrifood sector, it represents between 17 and 32 per cent of all global human-induced greenhouse gas (GHG) emissions (Greenpeace 2008).[1] Add to this the amount of GHG produced by food processing, distribution, storage, preparation and disposal, and the overall impact of food production and consumption on climate change becomes visible. At the same time, agricultural production is highly vulnerable to climate change, as the latter has been linked to the likely loss of huge areas of productive land. Accordingly, it seems pertinent to investigate the interaction between the global agrifood system and climate change in more detail.

In order to understand this interaction and identify the most crucial aspects for this analysis, one has to pay attention to the changes that have taken place in the global agrifood system in the last decades. First, due to processes of capital concentration the agrifood sector is dominated today by transnational corporate actors, which have established oligopolies in almost all segments of the system. Second, in conjunction with this expansion in economic power, corporate actors have assumed political power to a previously unknown degree. One way in which they exercise this power is through the creation and implementation of private voluntary

Business and climate policy: The potentials and pitfalls of private voluntary programs, Ronit (ed.), United Nations University Press, 2012, ISBN 978-92-808-1214-5

programs, i.e. institutions defining rules and standards for the global agrifood system.[2] While these rules and norms are directed most immediately at business, they hold pertinent implications for all stakeholders, ranging from farmers and producers of farm inputs to consumers in the global agrifood system. Accordingly, they deserve our urgent attention, because these private voluntary programs have dramatically expanded in number and reach. As a result, private retail food programs and standards have assumed a highly influential position in this setting and have become a dominant structural force in global agrifood governance in the last decade.

Given that private retail food programs play such an important role in global agrifood governance today, they appear to be a particularly promising avenue for investigating the interaction between the global agrifood system and climate change.[3] After all, private retail food programs may offer a singularly effective point of intervention in reducing the GHG emissions associated with global food production and consumption. Retailers have already demonstrated their power to set standards for food production and processing with a global reach. Due to the oligopolistic nature of the retail food market, these *de jure* voluntary standards easily assume a de facto mandatory nature. Moreover, the organizational structures for the implementation and monitoring of such rules and standards are already in place. In other words, food retail corporations today have the power and the instruments available to set, implement, monitor and enforce private voluntary programs targeting the climate change impacts of the agrifood sector, if they want to do so. The following questions, therefore, need to be asked: To what extent do the rules and standards set by private retail food programs address climate change issues? To what extent are they effective in addressing this issue? And what are the determinants of the extent and effectiveness of private voluntary programs in this respect?

A first glance at the empirical evidence suggests that retail food corporations are indeed active with respect to the issue of climate change. Walmart, for instance, launched its "Sustainability 360" campaign in 2005, in which it announced a target of obtaining 100 per cent of its energy from renewable sources as one of three core goals (Walmart 2009). Yet, the interaction between private voluntary programs and the climate change implications of the global agrifood system has not been systematically investigated to a sufficient extent. To date, the link between climate change, private governance and agrifood remains a "black box" in academic research and on the political agenda.

This chapter aims to provide a first set of answers to the questions raised above, building on, extending and applying the framework laid out in Chapter 1. It analyzes a range of relevant private voluntary programs in the retail food sector, assesses the extent to which they address the issue of climate change and their likely transparency and effectiveness, and attempts to provide explanations for these developments. The chapter will show that some of the largest retailers are indeed active in this area, while many of the group programs, such as the Global Partnership for Good Agricultural Practice (GlobalGAP), International Food Standard (IFS) and the Global Food Safety Initiative (GFSI), are not. Using a broad range of criteria, the chapter systematically evaluates and compares the different initiatives. Moreover, it critically reflects on the programs' activities and their range and promise of effectiveness.

Finally, in terms of explaining the coverage and stringency of private voluntary programs in the food retail sector and the issue of climate change, the chapter will argue that the reasons for the lack of effectiveness and in other cases the complete neglect of climate change issues in general are twofold. First, the link between the agrifood sector and climate change is not immediately visible to the public/consumer and therefore there is a corresponding lack of pressure on the actors creating and implementing the respective private governance institutions to integrate climate change objectives. Second, addressing the most important sources of climate change in the agrifood sector would imply fundamental changes in its overall design and functioning rather than the regulation and optimization of certain specific processes. Existing private governance institutions, however, rarely (if ever) target such fundamental changes.

The chapter is structured as follows. The next section delineates developments in the role of business actors in general and retail food corporations in particular with regard to the global agrifood system and global agrifood governance. A third section further develops the analytical framework presented in Chapter 1 and lays out a theoretical model for explaining the coverage and effectiveness of private voluntary programs, identifying the various incentive structures for firms to build and join programs. The following section then illustrates the application of the theoretical framework. Due to the preliminary nature of the assessment as well as the recent[4] creation of many of the relevant programs, this illustration first presents the private voluntary programs by retail food corporations and relevant business groups and their goals and activities with respect to climate change, and in a second step links the insights gained on the programs back to the theoretical framework, thereby providing a first set of ideas for explaining their coverage and effectiveness. The chapter concludes with a short summary and outlook.

Private actors in agrifood governance

Transnational actors have been playing a pivotal role in the global agrifood system for a while now. Globalization, with the associated trends of

Table 5.1 The market power of transnational corporations in the agricultural sector.

Product	Share in global exports markets by 3–6 of the largest TNCs in agricultural sector
Wheat	80–90%
Corn	85–90%
Sugar	60%
Coffee	85–90%
Rice	70%
Cocoa	85%
Tea	80%
Bananas	70–75%
Wood	90%
Cotton	85–90%
Pelts, furs and skins	25%
Tobacco	85–90%
Caoutchouc	70–75%
Jute and jute products	85–90%

Source: Enquete Commission Globalization of the World Economy, Deutscher Bundestag, 2002.

the liberalization of trade and capital flows, has fostered the development of business actors of an enormous size and reach in the agrifood system (Bonanno, Busch, Friedland, Gouveia and Mingione 1994). The underlying processes of capital concentration, in turn, have meant the development of oligopolies in which just a handful (or less) of these actors control a large share of the market at almost every stage of the supply chain (see Table 5.1).

While such oligopolies used to be more prominent in the production of input for the agricultural end of the food system as well as in the food processing stage, the process of capital concentration also became very visible at the retail end of the supply chain, in the last decade. Here, this process was facilitated and strengthened by new technological and logistical developments allowing better control from farm to fork, as well as competition based on quality aspects of food products rather than just price (Burch and Lawrence 2005; Konefal, Mascarenhas and Hatanaka 2005). Today, we can recognize ten large internationally operating retail chains with aggressive expansion strategies, and experts predict further processes of capital concentration in this market (Dixon 2007).

Global reach, capital concentration and the existence of oligopolies only tell one part of the story regarding the nature of today's global agrifood system, however. Another important aspect is the dramatic expansion in private voluntary programs in this agrifood system in the last decades.[5] Via a diversity of self-regulatory measures, transnational corporations play an enormous role in the authoritative allocation of values, i.e. the politics, of the global agrifood system today. Private voluntary programs now exist in almost all spheres and sectors of the global agrifood system, reaching from the production of agricultural inputs, food products or biofuel to traceability schemes and food safety standards created and implemented by actors at the retail end of the supply chain.

This development is all the more noteworthy, as it used to be corporate actors that were the prime focus of agrifood governance, whose conduct was deemed to be in need of monitoring and regulation. And not surprisingly so. After all, agricultural production and food provision are highly sensitive policy fields, at the core of any government's task to ensure the well-being of its population.

Today, however, these transnationals have become important subjects rather than just objects of global agrifood governance (Clapp and Fuchs 2009; Graz and Noelke 2007). Similar developments have been documented for other policy fields. The literature identifies a range of reasons for this trend. Optimistic observers argue that private actors have had to fill an existing void in public governance in order to be able to function in this global market (Biedermann 2007). Critical observers, however, point to attempts to increase market shares and rents, the wish to preempt public regulation, and a supportive neoliberal Zeitgeist as motors behind this development (Drache 2001; Gibson 1999). If we apply these perspectives to the agrifood sector, we arrive at very different expectations regarding the impact of private governance on the sustainability of the global agrifood system, of course, as well as assessments of the democratic legitimacy of private agrifood governance (Porter and Ronit 2010).

Within private voluntary programs in the agrifood sector, retail programs play a special role, as they are able to shape incentive structures along the supply chain. Retail corporations have dramatically increased their structural power in the past decade due to the combination of the process of capital concentration, pointed out above, and their advantageous position in the market, i.e. their proximity to the consumer (Fuchs, Kalfagianni and Arentsen 2009). Within just a few years, private standards and certification systems have become powerful gatekeepers for access to the global market and economic opportunities within it. While private governance is *de jure* voluntary, private retail standards are one of the clearest examples that, de facto, it may often not be voluntary at all. Private retail food standards and certification systems define criteria for food production and processing, with an emphasis on quality and food safety issues and some attention also being paid to environmental and social aspects. Suppliers need to be able to implement the rules and standards and document their compliance if they want to be able to sell their products in the global market, i.e. be part of the supply chain of one of the major retail corporations.[6]

As a consequence of the enormous power which private retail food standards exercise in today's global agrifood system, this investigation into the implications of private voluntary programs in the food sector for climate change concentrates on them. It examines whether, to what extent and how these standards address the issue of climate change. Moreover, it attempts to provide a first set of explanations for such a focus (or lack of it) as well as an assessment of the expected impact.

The coverage and effectiveness of private voluntary programs

In this section, we present a theoretical framework to analyze the behavior of firms and industries to identify determinants of the evolution of private voluntary programs and examine the questions raised above. This framework includes an input and an output dimension. Input refers to whether programs address certain issues and include certain members, as well as including questions such as the substantive stringency of a standard. Output refers to the effectiveness of programs and whether actual changes in business conduct are achieved in the course of the implementation of the agreed private voluntary program. This actual behavior is a function of the agreed standards, i.e. of the rules and membership structure, of course, but also of existing incentives and opportunities to comply with the standards, to outperform them, or to fail to comply.

In following this structure, we further develop and apply the four basic elements of private voluntary programs delineated in Chapter 1: designing, joining, monitoring and complying. Thus, we treat aspects of designing and joining in the question of the determinants of input. It is difficult to strictly differentiate between these elements due to the specific nature of our field of inquiry, in which designing has played a much more important role than joining, and these aspects have so far been more developed than monitoring and complying. Because of the explorative character of the research and its focus on relatively recent developments, how and why companies create and install different standards in the first place is of primary interest. Individual initiatives dominate the field at this point, and those group standards that do exist have been developed jointly by the large retailers. This, in turn, implies that smaller actors along the supply chain have limited choices when it comes to the question of joining. Accordingly, the question of joining is not that relevant in our analysis. In terms of issue coverage, in particular, the important choices are being made at the design stage.

We treat aspects of monitoring and complying in the section on output. The question of the extent to which actual conduct is being monitored, by whom, and with what consequences, i.e. whether there are sanctions or public reporting on compliance failures, strongly influences the likelihood of consequences, and thereby the achievements of private voluntary programs. Of course, there is also an influence of monitoring and sanctioning provisions on the willingness of an actor to join a private voluntary program. Likewise, there is an interaction between what the actors creating a program are willing to subscribe to and the performance of programs. Admittedly, these interactions need to be further developed in our analysis in the future.

Input: Designing and joining

Output of private voluntary programs, i.e. the determined standard, can be illustrated as the consequence of cost–benefit analyses of participating business actors. Institutional as well as material goods are usually characterized by a bulk of property rights which show different characteristics regarding their divisibility and usability (Fuchs 2003). Governance institutions, which at first sight serve the provision of public goods, have private benefits as well. If a company can reduce the risk of a scandal resulting from ecological and economic harm by, for example, introducing a system fostering food safety or reducing the climate impact, then this system will generate not only better food safety or less environmental pollution as a good for society but also a private benefit for the company. Likewise, the basic improvement of a company's image as a provider of high-quality products and as a good steward to the environment, the opportunity to gain higher prices, or the prevention of expensive governmental regulation add to this private benefit of private rule-setting.

For business purposes, it is reasonable to invest in and join private voluntary programs as long as the private benefits of the relevant program exceed its private costs. This issue is also dealt with in the context of club theory, which is developed and applied in Chapter 3 by Hsueh and Prakash. Figure 5.1 shows this cost–benefit function for two companies, A and B. The cost of investment in the program (C_A and C_B) increases with the stringency (S) of a standard to an increasing degree. "Stringency" here means substantive performance. One can differentiate between designed climate standards according to whether they define only process standards, or performance standards as well, for instance, and in terms of the ambition of the performance criteria. While first improvements generally can be attained with relatively low investments, the marginal costs of improvement increase with rising standard stringency. The position of the cost function depends on technological and organizational characteristics

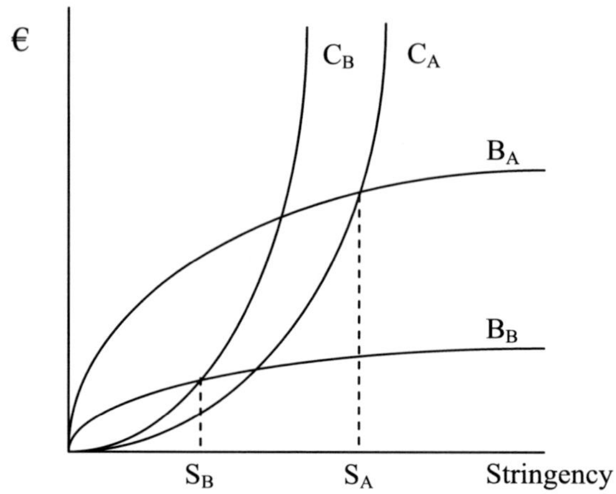

Figure 5.1 Investment decisions regarding private voluntary programs.

of the company and can vary, for example, due to the existence of economies of scale or differences in the know-how of companies. In addition, the private benefit of investments in private voluntary programs, for example in the form of a reduction in the risk of scandals, increases with such investments (B_A and B_B), but to a decreasing degree. While first investments can be expected to deliver substantial improvements regarding the risk potential, further investments bring smaller reductions in risk.

The different cost and benefit functions of companies follow from the fact that not all of them have the same opportunity of transforming governance investments into corresponding economic gains, as we shall see later. Companies which are more vulnerable to civil society campaigns or new public regulation due to their company and brand image, sector characteristics, cultural origin, or proximity to consumers, or which are more sensitively hit by political consumption strategies, can, in principle, benefit more from private voluntary programs than companies for which the various factors do not apply. The resources of civil society organizations for monitoring business conduct are limited. Also, the boundaries of public receptiveness and reactivity regarding the revelation of "scandals" show that only the companies which, so to speak, stand in the front row of public perception feel public pressure to its full extent. Corporations especially, which are of pivotal interest in the present research inquiry, are in close contact with consumers and are therefore very vulnerable to consumer attitudes towards the company.[7]

As emphasized in Chapter 1 and elsewhere in this book, general and sector-specific knowledge and values play a role. Indeed, the presence or lack of legal and accounting expertise may exercise an influence on a company's assessment of the costs and benefits of investments in a private voluntary program. Similarly, the benefit function's position will be influenced by deliberate and unconscious decisions regarding the valuation of risks and opportunities as well. In this respect, top management's normative perspective and their connected social and intra-corporate learning processes also have an effect (Cutler, Haufler and Porter 1999; Nash and Ehrenfeld 1997).

Next to company characteristics, the problem characteristics of climate change will have an influence on cost–benefit functions. Thus, the extent of a problem, filtered through available information and perceptions, as well as the likely impact of the private voluntary program on this problem, will affect a company's willingness to invest in the program. Likewise, public awareness of the existence of a problem and its relationship to the company's activities, which influences the likelihood of public or consumer pressure, is going to have an effect. Indeed, companies may gain more from investing in private voluntary programs targeting marginal issues or issues in a superficial way, if these issues or superficial measures have or can easily gain the attention of the public. Similarly, the management's norms and beliefs regarding the particular problem can play a role.

Finally, the formal involvement of civil society or public actors influences the utility functions of participating businesses as well. The involvement of civil society may lead to higher levels of governance investments, when designing, implementing and enforcing rules, for instance. This effect can be explained easily from the benefit point of view since civil society's participation usually implicates higher credibility and, therefore, higher private benefits. The same applies to participating public actors, admittedly to a somewhat lesser extent due to the – at least in some countries – growing concern about a potential capture of the state.[8]

On the cost side, technological and structural factors are likely to play a role in climate change policy. Thus, the cost function is likely to be placed lower for issues that can be solved with available and affordable technological changes or small organizational efforts, which are really critical issues in the field of climate change policy. For technologies that need to be developed first (in which case there is also a risk of development failure) or for fundamental changes affecting the core of a company's activities and existence, cost functions may be deterringly high. Similarly, company as well industry profiles differ with respect to their exposure to certain problems, of course. Thus, a company in a highly energy-intensive sector may face higher costs with respect to energy savings measures than a company in a sector with low energy intensity.

In sum, the following major factors can be expected to exercise an influence on an actor's utility function regarding investments in private voluntary programs: (1) factors influencing its visibility and perceived pressure: the size of the company, country of origin, host country, brand/ business and sector history and image, range of products, problem characteristics; (2) factors influencing an actor's normative position: values of the executive board, affiliation in business associations, problem characteristics; (3) additional factors influencing the opportunity for private profit: competitiveness of the environment, provision of legitimacy due to civil society and/or state involvement; (4) factors influencing the cost function specifically: extent and nature of necessary changes, availability and affordability of technological and organizational solutions, and product and process characteristics with respect to the problem in question.

The perspective on cost–benefit functions of business actors demonstrates that a company will prefer a more stringent rule-setting, the higher the privatizable benefit of the rules and the lower the cost of required investments. In Figure 5.1, company A has relatively higher opportunities to privatize the benefits of governance institutions and a relatively lower cost function than company B. Accordingly, investments in such a governance institution can be profitable for company A up to point S_A, while investments for company B are only lucrative up to point S_B. However from an economic point of view, the ideal point of investment would be attained earlier in each case, namely, at the point of maximum net benefit, which is shown in Figure 5.2 for company A with S_{A*} (and, of course, can be equally determined for company B).[9]

The different cost–benefit functions highlight that for private voluntary programs with more than a single member company, a compromise will frequently be needed. In other words, for group standards (e.g. sectoral

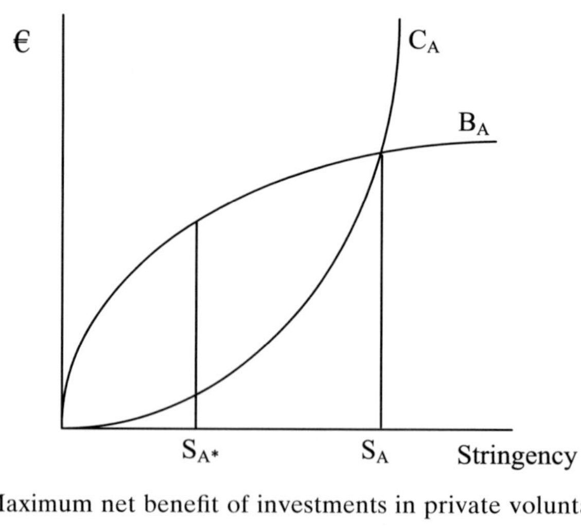

Figure 5.2 Maximum net benefit of investments in private voluntary programs.

standards) rather than individual private voluntary programs, the process of negotiation among the different actors needs to be considered as well. The impetus for the development of such a group standard may be the experience with public or civil society pressures, the existence of similar normative predispositions of the management, or a need to solve common problems (besides pressure by other actors). Likewise, such a process may be initiated by an external agenda-setter, such as a business association, civil society organization or effective and recognized body outside the industry in general, as discussed in Chapter 1. In addition, a company may itself actively recruit others to create a joint program. A precondition for the successful development of a group standard, however, would seem to be a minimum of common interests among the powerful actors in the group. In consequence, we can add a fourth point to the list of determinants of the output of private voluntary programs identified above, namely, factors influencing the development of group standards: minimum of common interests and the distribution of power among actors in the group. We will see later how this factor affects group programs in the agrifood sector.

In our example, the stringency of the agreed standard, or the output of the private voluntary program between company A and B, would lie somewhere between S_{A*} and S_{B*}, including both end points ($S_{B*} \leq S_{AB} \leq S_{A*}$). In other words, there are three possibilities: an agreement on the minimum standard ($S_{B*} = S_{AB}$), a compromise on a standard between both ideal points ($S_{B*} < S_{AB} < S_{A*}$), or an agreement on the maximum standard ($S_{A*} = S_{AB}$).

At the same time, this leads to possible strategies for external actors to influence input into private voluntary programs. Thus, civil society organizations or a "shadow of hierarchy" created by public actors can increase the cost of failure for business actors, and, therefore, move the companies' benefit functions to the upper right on the graph. In this context, a focus by civil society organizations or public actors on companies of type B suggests itself, since a movement of their benefit functions would most clearly improve the agreed standard.[10]

Output: Monitoring and complying

The three above-mentioned possibilities of input contain different implications for the implementation of the standard and, therefore, the output or effectiveness of private voluntary programs. In the case of an agreement on minimum standard (S_{B*}), the incentive arises for company A to invest beyond the standard. Such investments by A will be smaller, however, as a softer standard cannot be communicated with as much gain. When there is a compromise between the positions, incentives of non-

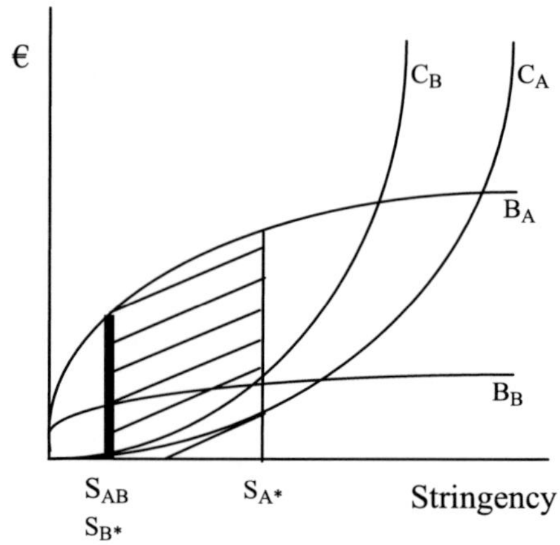

Figure 5.3a Investment decisions and performance bands I ($S_{B*} = S_{AB}$).

compliance arise for company B, besides the incentives for A to exceed the standard. For the third model, the agreement on maximum standard, the incentive for company B not to comply with the standards is the largest. In this respect, the effectiveness of a rule should be thought of as a band of performance around the agreed standard S_{AB}, which allows for both leaders and laggards (Figure 5.3).

The direction of this band will depend on a number of factors. If the actors agree on a minimum standard ($S_{B*} = S_{AB}$), the band will be to the right of the agreed point of investment (Figure 5.3a). When there is a compromise between both preferred positions ($S_{B*} < S_{AB} < S_{A*}$), it will be both above and below the agreed standard (Figure 5.3b), and when a high standard is agreed on, the band will be to the left of the point of investment (Figure 5.3c). The spread of the band below the standard will be influenced by the existence of monitoring, sanctioning and reporting mechanisms, among others. These will influence the negotiations on the stringency of the standard already, as company B will refuse to accept a high standard that is coupled with strong opportunities for monitoring and sanctioning. Thus, the actual effectiveness of private voluntary programs can clearly exceed the agreed standard or clearly fall below it.

When can one expect which form and direction of a performance band? Generally, tight performance bands are only likely within small groups of relatively homogeneous actors, and the agrifood sector is highly oligopolistic. However, as soon as we talk about more encompassing rules including a large group of firms, a relatively extended stretch should be expected.

In addition, the factors influencing the direction of the band of performance can be identified. The agreed climate standard will depend on

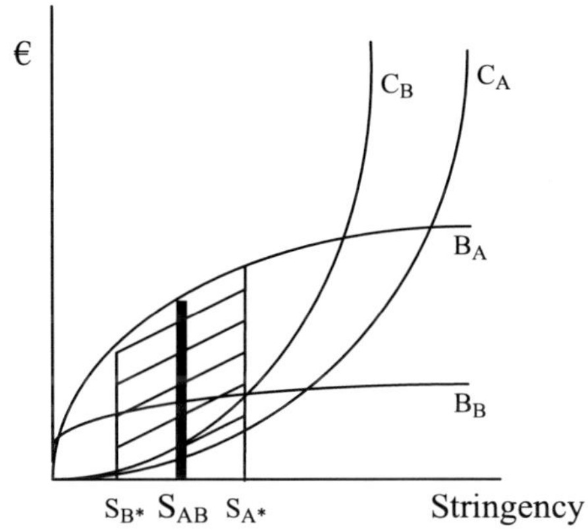

Figure 5.3b Investment decisions and performance bands II ($S_{B*} < S_{AB} < S_{A*}$).

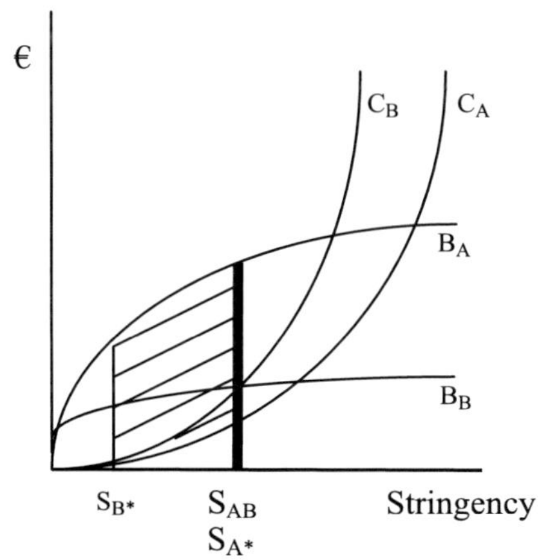

Figure 5.3c Investment decisions and performance bands III ($S_{A*} = S_{AB}$).

the power distribution among companies, control and penalty possibilities, as well as on costs associated with non-compliance of individual companies. Thereby, it is assumed that the case of standard agreement is rare at the upper end of the performance spectrum. A highly ambitious standard which allows unsanctioned defection by laggards only makes sense for leaders if neither the standard's reputation is damaged nor their own. Such damage could result, after all, in the case when failure of standard performance by laggards becomes public. A highly ambitious standard that sanctions defectors, i.e. with efficient control and sanctioning mechanisms, however, will hardly be acceptable for mass-market companies, and, inasmuch, may only exist in the case of a strong asymmetric power distribution in favor of companies with higher benefit functions.

Moreover, one can expect that the less relative influence companies with high benefit functions have in negotiations, and the more monitoring and sanctioning mechanisms are created, the closer to the left end of the performance band a group standard will be. Such sanctioning mechanisms will be needed more if the group is large and heterogeneous, i.e. the bigger are the collective action problems within the group and the incentive to defect. In such situations, one is more likely to encounter rather low standards, or standards which secure the minimum performance and, simultaneously, give leeway to the better performance of businesses, and foster the expansion of the performance band towards the right through incentive structures and learning processes. Hence, one can derive a minimum of program effectiveness from the standard's stringency in such situations and, simultaneously, consider a further potential positive momentum from learning processes and incentives to outperform the standard. One should not infer standard effectiveness only from the good performance of individual leader companies, however. In other words, the usefulness of individual case studies or even surveys of a select range of companies (such as the largest) or companies in the "front row" is limited here.

Common rules and standards within the midfield of the performance band should be expected as a compromise solution which allows a certain range of actual conduct in both directions. However, leaders should have an interest in standards not being undercut too often and explicitly, and, thereby, losing their private benefit. Such standards can be found in situations in which informal control and penalty mechanisms exist due to group characteristics, and collective action problems remain manageable. The standard's stringency itself provides an indication of its average program effectiveness in such cases. However, due to their frequent focus on leaders or laggards, the value of case studies on individual companies or a select number of companies is limited here as well. Thus, empirical studies would need to analyze the full range of relevant companies or a random sample.

In sum, the following factors can be expected to determine the extent to which a program actually changes the conduct of a business and its pattern of complying, i.e. the output of a private voluntary program: (1) factors influencing the collective action problems: size and heterogeneity of target group; (2) power asymmetries among the actors involved; and (3) factors influencing the likelihood of defection: monitoring and sanctioning opportunities, public reporting of compliance failures.

Retailers' climate change governance in the agrifood sector

This section provides an overview, assessment and first set of explanations of private retail food programs that address climate change issues and their specifications. It first lays out the activities of the ten largest food retail corporations, and subsequently describes efforts by retailer groups and associations. In our approach, we use four categories based on the theoretical framework outlined above to measure the stringency (input) and the effectiveness (output) of private retail programs in tackling climate change: designing targets, partners and standards, transparency and reporting, and governance mechanisms for monitoring (see Table 5.2). The first two categories emphasize the input side of the process, while the latter relate more strongly to output. In a second step, the information gained on the programs will be related back to the theoretical framework laid out above, in order to provide a first set of ideas on the influence of the potential determinants of program design and effectiveness.

We focus on the world's ten leading food retailers (Walmart, Carrefour, Tesco, the Metro Group, Kroger, Auchan, Rewe, Aldi,[11] Lidl[12] and Costco) and the most relevant groups (British Retail Consortium, BRC; the European Retail Round Table, ERRT; the Global Food Safety Initiative, GFSI; Global Good Agricultural Practice, GlobalGAP; and the International Food Standard, IFS), because of their importance in current developments in the global agrifood system and global agrifood governance. These corporations and associations are the dominant actors in the food sector, responsible for the processes and fundamental changes which were described earlier. Thus, we are able to avoid selecting our cases on the basis of the dependent variable and can ensure a comprehensive picture of the relevant actors. In other words, we are able to point out which of these retailers and groups are neglecting the issue of climate change.

Importantly, the role of food in the product portfolio varies for the retailers and groups analyzed. For Walmart, Carrefour, Kroger and the Metro Group, other products such as toys, electronics, clothing or jewelry are at least as much a core focus of their business activities as food. Likewise the BRC and ERRT, two of the groups whose private voluntary programs we consider, do not focus on food retail specifically, in contrast for instance to the GFSI, GlobalGAP and the IFS. We will reconsider this difference in portfolio, when reflecting on the activities by the various retailers and groups with respect to climate change.

Designing targets

This first category is closely associated with the inquiry into issues of designing private voluntary programs. As decisions about design tend to

Table 5.2 Private voluntary programs in the agrifood sector and climate change.

	Designing targets				Partners and standards		Transparency and reporting			Monitoring
	Issue of climate change	Targets	Renewable energy	Supplier	Partnerships	Standards	Information	Reports	Label	Monitoring
Walmart	++	++	+	++	++	++	++	++	–	++
Tesco	+	+	++	+	++	++	++	+	++	+
Carrefour	++	++	–	+	++	+	++	+	+	+
Metro	+	+	+	–	+	–/+	–/+	+	–	–
Kroger	+	++	–	–/+	–	–	–/+	+	–	–/+
Rewe	+	+	++	–	+	+	++	+	–/+	–/+
Auchan	+	–	+	–	+	–/+	+	+	+	–
Aldi South	+	–	+	–	–	–	–/+	–	–	–
Lidl	–/+	–	–/+	–	–/+	–	–/+	–	–	–
Costco	+/–	–	–	–	–	–/+	–	–	–	–
Aldi North	–	–	–	–	–	–	–	–	–	–

– the criterion has not been met; –/+ the criterion has been partly met; + the criterion has been met; ++ the criterion has been more than met.

have consequences for joining, our analysis also indirectly relates to that issue, of course. "Designing targets" examines the objectives and levels of stringency determined in the explicit design of private voluntary programs. It asks whether the programs address the relevant questions, in our case climate change. It also explores to what extent specific targets have been set, for instance in terms of emissions reductions. In addition, the design of private voluntary programs can be assessed in terms of investments in related issues, such as renewable energy. Finally, the range of processes and actors addressed by private voluntary programs has important implications for the potential impact of these programs. Accordingly, this section asks to what extent the identified programs target the retailers' supply chains.

When looking at these four issues, we find a wide variety in program design. Walmart and Tesco explicitly address all four aspects in their programs. Walmart, for example, has initiated a long-term program, in which it spells out its aim "to reach a day where all of our energy comes from a renewable source" (Walmart 2009: 6). The corporation created the "Sustainability 360" approach, in which it attempts to integrate its associates, suppliers, communities and customers in its climate policy (Walmart 2008a). Walmart's "sustainability goals" are: to source 100 per cent of energy from renewable sources, create zero waste, and sell products that sustain the world's resources and environment. To achieve these goals, the company has set several benchmarks to improve the energy efficiency of its stores and trucking fleet. While sounding quite ambitious, Walmart's efforts contain a major weakness: the program fails to set a date by which the goal "Sustainability 360" should be realized.

Tesco has initiated a climate change program with three main parts: the company aims to reduce its own direct footprint; it works with its supply chains and partners to reduce emissions more broadly; and it wishes to lead a revolution in green consumption (Tesco 2009). Specifically, Tesco has started to set targets which address the reduction of GHG emissions in its buildings and its goods. Here, it wants to halve emissions from existing buildings by 2020, to halve distribution emissions of each case of goods delivered by 2012, and to halve emissions from new stores by 2020 (all against a baseline of 2006). To achieve its goals, in 2007 the company invested £100 million in a Sustainability Technology Fund to support large-scale world-wide carbon reduction technologies in its stores, distribution centers and supply chains.

Carrefour and Rewe also perform relatively well in terms of designing targets, each addressing three of the four aspects. As part of the relevant measures, Rewe, for instance, has announced an aim to cut its greenhouse gas emission by 30 per cent by 2015. Moreover, Rewe claims that it is obtaining nearly 100 per cent of its energy demand in Germany from renewable energies (Rewe 2009). Carrefour made a commitment in 2004 to reduce its energy consumption by 20 per cent per square meter of sales area by 2020. To achieve its target, the Carrefour Group initiated an Energy Management System (EMS) project – a theme also referred to by Clapp and Thistlethwaite in Chapter 2 – to enable it to develop telemetry and remote control of equipment (Carrefour Group 2009).

The programs of Kroger and the Metro Group take a middle position. They are concerned with climate policy issues, but the various programs are not as detailed as those of the four previously mentioned corporations. The North American food retailer Kroger made a commitment for change in three areas[13] in which it plans to reduce its environmental footprint: energy conservation, emissions reduction and waste reduction (Kroger 2009). Ambitiously, Kroger specifies a goal of reducing the overall energy consumption in stores by 30 per cent by 2010, using 2000 as a base. Kroger states that "it has worked aggressively in all areas of [its] business to reduce energy consumption" (Kroger 2009: 8). It claims to already have reduced its overall energy consumption by more than 22 per cent – or 1.6 billion kilowatt hours or 1 million metric tons of GHG emissions – since 2000 (Kroger 2009). However, Kroger fails to comprehensively address suppliers or the question of renewable energy sources. The Metro Group has mentioned specific objectives in its reports since 2007. It has committed to decreasing its emissions in the period from 2006 to 2015 by 15 per cent per square meter of sales area. To achieve its objectives, the Metro Group publicly promotes its goals to change its own as well as other actors' behavior (Metro Group 2008).

Efforts by Auchan and Aldi South show even less specificity, thereby raising questions regarding the seriousness of the companies' efforts. Auchan does not formulate precise goals, but commits itself to reducing its energy consumption, carbon emissions and emissions from transport. In addition, it has announced plans to invest in new technologies, innovations and renewable energy to reduce its impact on the climate (Auchan 2009). Aldi South has committed itself to three types of climate activities: saving energy in CO_2-intensive production mechanisms (fossil fuels), obtaining energy from renewable sources, and protecting the climate (Aldi South 2009a, 2009b, 2009c). However, there is no commitment to explicitly defined goals. Moreover, it is noteworthy that Aldi South only addresses these issues on its German website.

Finally, Lidl, Aldi North and Costco rarely mention the issue of climate change. If we look at the design of the environmental program of the three companies, we see that these corporations have no commitments to CO_2 reductions or related environmental issues.

The specificity, stringency and comprehensiveness of the targets designed into the private voluntary programs by retail food corporations vary widely, then. This is all the more noteworthy, as even the best programs still tend to have considerable weaknesses. The failure to set target dates, for instance, means that it will be difficult for stakeholders to hold the company to any of its promises. More importantly, most of the targets specified for emissions reductions, for instance, focus on the easy aspects of food retailing, such as store lighting, cooling or the efficiency of the truck fleet. The more difficult aspects of the climate change impact of the agrifood system, such as the overall distances traveled by products sourced and distributed in a global system, or the rise in the average carbon footprint of food products resulting from the ongoing industrialization of production methods, or an increasing share of meat products in diets, for instance, are rarely addressed.

Partners and standards

The second category inquires into public or civil society partners involved in the design and implementation of food retailers' private voluntary programs as well as international or national standards facilitating the accessibility and comparability of the programs.[14] Both aspects are primarily a question of program design (and thereby joining), but also relate to output issues in the form of complying, as these external actors and standards potentially increase the likelihood of compliance. In terms of design, inclusion of partners and standards able to allow for benchmarking suggests a greater probability of stringency of programs.

Public–private partnerships and private–private partnerships (e.g. with NGOs) are part of the policies of Carrefour, Tesco and Walmart. In 2002 and 2004, Carrefour conducted a carbon assessment of its stores in partnership with the ADEME (French Environment and Energy Management Agency). In addition, it joined the Supply Chain Leadership Collaboration in 2008 "to raise awareness ... of the effect of CO_2 emissions and general climate change" (Carrefour Group 2009). The public institution supports the company to calculate the ecological footprint of the products provided by the retailer. Like Carrefour, Tesco has different partners on different sectoral levels. To decrease its carbon footprint in the distribution sector for example, it works with the Institute of Grocery Distribution, a UK charity, which informs and educates people who work in grocery stores, on how to implement best practices for reducing CO_2 emissions (Tesco 2009). Simultaneously, Tesco is working with the University of Manchester and Defra (Department for Environment, Food and Rural Affairs) to gain support to reduce the environmental impact of the company. Walmart and the Clinton Climate Initiative, in turn, announced a partnership on the US Mayors' Climate Protection Summit in 2007 (Walmart, no date). This partnership plans to support the introduction of environmentally friendly technologies, such as energy-efficient building materials and systems, and to explore ways to use their purchasing resources to reduce prices on sustainable technologies (Walmart 2009a). In addition to this initiative, Walmart has created a Food and Agriculture Network, a coalition of buyers, suppliers, civil society organizations and academics, to decrease the environmental impact of food miles, water use and degradation, and packaging, and to improve energy efficiency (Walmart 2009a).

Some of the other retailers also work with public or civil society partners, but do so to a lesser extent or work with weaker partners. The Metro Group and Rewe specify that they cooperate with the German Oeko-Institut e.V. (Institute for Applied Ecology), an environmental think tank, in their efforts. In addition, the energy provider Energie-Handels-Gesellschaft supports Rewe in obtaining its electricity from renewable sources, according to Rewe (Rewe 2009). As this partner seems to exist only for this purpose, however, its dependency on Rewe means that the partnership cannot necessarily be expected to increase program stringency. Auchan is also a partner of ADEME. Moreover, it states that its commitments to climate issues are based on the ideas of the Grenelle de l'environnement, a debate held in France in 2007 between the government, professional associations, private actors and civil society (Auchan 2009).

Lidl, Kroger, Aldi North and Aldi South do not provide any information about partnerships.

If we address the question of standards, we find a substantial difference among the different private voluntary programs of the food retailers. While some retailers mention a range of standards or benchmarks, others do not address any of them. Tesco, Carrefour, Costco and Rewe, for instance, use the Greenhouse Gas Protocol[15] to measure their CO_2 emissions. As a supporting instrument, Tesco, Carrefour, the Metro Group and Rewe use the ISO Standards 14040 and 14044, also discussed in Chapter 2, to calculate their emissions. In addition, Tesco and Walmart have been cooperating with the Carbon Disclosure Project since 2007 and publish their carbon footprints through it. Moreover, since 2002, Carrefour and Walmart have published their sustainability reports in line with the Global Reporting Initiative (GRI) (Carrefour Group 2005; Walmart 2009). In contrast, Kroger, Lidl and Aldi South do not employ any standards that would allow a comparative assessment of their programs.

To sum up, we can see that partners and standards serve to differentiate between the design of the corporations' voluntary programs with respect to climate change. Particularly noticeable is that the biggest retailers (Walmart, Carrefour and Tesco) appear to be quite ambitious in joining standards and maintaining partnerships. Clearly, the internal weakness of many of the standards employed means that they cannot ensure the stringency nor effectiveness of private voluntary programs by food retailers. Moreover, the partnerships vary in their extent and in the power relationship between the partners. Still, such partnerships and standards allow the characterization and assessment of the various programs on an additional dimension.

Transparency and reporting

The third category discusses various aspects of providing information and transparency to external stakeholders. It is thus related to designing and joining, and also directly to complying, as the potential for external assessment is likely to increase pressure for compliance. Thereby, this category also moves the aspect of program effectiveness (output) more into the center of attention. To assess transparency and reporting, we look at how much information the companies provide and how transparent the access to this information is. In addition, we explore whether the company publishes a sustainability report. Both aspects only provide a basic indication of program performance, of course, as information in sustainability reports is notorious for its vagueness and promotional character. Nevertheless, such information needs to exist if external stakeholders are to have any chance of gaining an impression of a program. Finally, we investigate whether the programs attempt to capture the environmental impact of products through a carbon label or similar instrument. Importantly, such instruments, specifically carbon labels, not only provide information to consumers as external stakeholders; they also ensure specificity in the provision of information and allow for a demand for change, which renders them an important element in the stringency and ambition designed into a program. While treating carbon labels in the category of "transparency and reporting", then, a clear link to the "designing targets" category should be noted.

With respect to transparency and reporting, we find that Tesco, Walmart and Carrefour provide a relatively broad range of information on their climate programs in the form of sustainability reports and other documents on their websites. Tesco and Carrefour have been publishing sustainability reports since 2002. Walmart only started in 2007, but since 2008 it has published an additional "Sustainability Progress Report", in which it presents an overview of the status of achievements regarding the company's environmental objectives (Walmart 2009). Rewe and Auchan only recently started publishing sustainability reports (2009), while those of the Metro Group and Kroger have been in existence longer, but contain relatively less information. In addition, Rewe and the Metro Group provide a climate brochure. In comparison to these retail food corporations, Lidl, Aldi South, Aldi North and Costco provide only limited information on their activities with respect to climate change. None of these four publishes sustainability reports. The German food retailers Aldi South and Lidl provide some information on their German websites. No information is available on other country websites. Costco provides relevant information only in its annual financial report.

If we look at the question of labeling, we see that Tesco, Carrefour, Rewe and Auchan are the only transnational corporations to discuss the labeling of products in the design of their programs. Tesco was the first food retailer to introduce carbon-labeled products.[16] By the end of February 2009, Tesco had labeled 100 products in its range. Importantly, these products do not include meat products. Auchan, too, introduced product carbon labeling to increase the environmental awareness of its customers. In 2008, Rewe started a pilot project to create and implement a carbon label for selected products. In the coming years, Rewe plans to increase the number of labeled products (Rewe 2009). In comparison to the corporations above, Carrefour argues that "by focusing only on greenhouse gas emissions, it obscures other environmental criteria, such as water, biodiversity and toxicity, which may be of paramount importance in the case of certain products" (Carrefour Group 2008: 11). Rather than pursuing the carbon label approach, therefore, Carrefour uses a life-cycle assessment to measure the environmental impacts of a product at "each stage of its life-cycle, from raw-material production to waste disposal" (Carrefour Group 2008: 12).

Information and labeling are important to characterize not only the design – the input dimension – but also, importantly, the effectiveness – the output dimension of private voluntary programs. Similar to our results for partnerships and standards, we find that only a few retailers seriously engage in the provision of information to external stakeholders. Again, the biggest retailers appear to be more ambitious in publishing and pursuing product labeling. Even those programs, however, still lack specificity and comprehensiveness in their climate change related reporting.

Monitoring

Our last category addresses the issue of monitoring, i.e. inquires into the internal governance mechanisms of the selected private voluntary programs. Again, internal monitoring mechanisms are a question of design (and joining), and thus relate to the input dimension. More importantly, however, they relate to the output function as monitoring is a precondition for even the most basic potential for internal pressures to ensure compliance. In other words, to answer the question whether the programs are likely to be effective, it is also necessary to analyze whether the design of the implemented standards and programs will be monitored and evaluated regularly. Therefore, we investigate whether information on evaluation, monitoring and control will be provided by the companies.

Walmart reports that it has initiated the so-called Walmart Sustainable Value Network to monitor the implementation of its objectives (Walmart 2008b). It is a cooperation between Walmart employees, civil society organizations, academics, politicians and suppliers, and has the mandate to monitor the targets in all business activities and report the performance to Walmart (Walmart 2008b). Carrefour and Tesco, in turn, have implemented "Key Performance Indicators" to evaluate their business activities. Carrefour has appointed KPMG to control its activities, while Tesco monitors its activities for itself.[17] Rewe points out that the company's partners (EHA and the Öko-Institut) control the achievements of its commitments. Interestingly, not only do Lidl, Aldi North, Aldi South and Costco not publish any information about monitoring, but the Metro Group and Auchan fail to do so as well. Kroger mentions monitoring, but here, too, the information stops at that and is thus insufficient (Kroger 2009).

In consequence, we find a similar level of variance in internal monitoring mechanisms for the programs investigated as we did for the other categories. The extent to which the world's biggest food retailers have built in monitoring mechanisms in their programs ranges from ambitious plans to a complete neglect of the issues, with the majority of companies falling somewhere in between. Importantly, even those with monitoring provisions do not specify remedies in the case of failure.

Group programs

When we turn to the group programs and check them against our four categories, we find that most programs, specifically the IFS and GFSI, appear to neglect the issue of climate change completely. They neither publish information on any kind of climate policy nor address the issue in their standards and principles. The GlobalGAP mentions on its homepage that its "standard is primarily designed to reassure consumers about how food is produced on the farm by minimising detrimental environmental impacts of farming operations [and] reducing the use of chemical inputs" (GlobalGAP 2009), but it is a challenge to find further information relating to environmental standards, and especially with respect to greenhouse gases. Therefore, we concentrate on the BRC and the ERRT.

If we look at the design we can say that the BRC and the ERRT seem to be relatively active in addressing the issue of climate change. The BRC has defined five overarching environmental goals, which are divided into several smaller targets and each includes climate-related aspects. For example, the first goal is to reduce the direct environmental impact of its retailers. As part of this goal, the BRC aims to reduce the emissions from its members' buildings by 15 per cent from 2005 levels by 2013, and energy-related transport CO_2 emissions from store deliveries by 15 per cent in the same period. The other goals comprise questions of the integration of suppliers or changing the behavior of customers. The ERRT initiated a pledge in 2008 to reduce its environmental impact by 20 per cent by 2020.[18] It prescribes a focus on energy efficiency where the ERRT, like the BRT, formulates five goals. First, members commit to reducing energy consumption per square meter of commercial premises by a minimum of 20 per cent by 2020 compared to base-year reference levels. Second, they wish to work towards a more ambitious target than the European Commission's target of sourcing 20 per cent renewable energy by 2020. Third, they identify and share examples of best practice in delivering energy-efficient solutions in the retail supply chain. Fourth, they attempt to investigate further ways of providing energy efficiency information for products they sell. Lastly, they wish to share knowledge about the most effective ways of communicating information on energy consumption and energy-saving behavior to consumers.

In terms of partners and standards, the BRC reports that it aims to engage partnerships to reduce the environmental impact of its members. The ERRT does not provide any information in this category.

Efforts to foster transparency and reporting about the programs of the two associations can be identified, however. Relevant information can be downloaded from their respective websites, which both provide press releases and brochures about their environmental commitments. To monitor the goals, the ERRT states that "the companies will report on progress through their annual reporting processes – for example, in their CSR reports, or specific energy efficiency reporting procedures, as appropriate" (ERRT 2008). In other words, ERRT reporting does not go beyond the reporting by the individual companies. The BRT asks its members to report their performance on their environmental goals relative to the 2005 baselines, as well as their plans for future action.

In terms of group programs, then, we can identify a couple of private voluntary programs addressing the issue of climate change by (food) retailers. These programs are in a very early stage, however, and lack ambition in terms of the inclusion of stakeholders and standards of comparability. The majority of associations and groups, however, are not active in this field.

Assessment and explanations

Having described the programs of the various retailers and groups, a critical reflection and explanatory assessment of the given climate change related activities and their likely effectiveness is necessary. Different aspects of input – in the form of designing and joining – and output – in the form of monitoring and complying – are covered. The comparative assessment shows that some food retailers, e.g. Walmart, Tesco and Carrefour, are relatively active in the climate change area, and their programs reflect some effort to reduce their carbon footprint. These actors have defined specific targets, work with civil society organizations, think-tanks or public actors in this context, and make their performance on the reduction of their carbon footprint relatively public and transparent. In addition, the BRC and the ERRT as associations have comparatively clear climate targets and provide information about their commitments to climate change. In contrast, the activities and the information published by German or US food retailers, e.g. Lidl, Aldi and Costco, or the GlobalGAP, are comparatively weak. Unsurprisingly, there is also a manifest lack of transparency. These activities engender little confidence in a reduction of the company's or group's carbon footprint. Finally, there are retailers and groups that do not have a program on climate change aspects at all, e.g. Aldi North, the GFSI and the IFS.

In general, the question of the accessibility of climate policy information from the respective retail corporations needs to be critically evaluated. With the exception of Lidl, Aldi and Costco, every retail chain considered here publishes an annual sustainability report.[19] Noticeably, Carrefour,[20] Tesco, Walmart and the Metro Group are the only retailers which provide the previous sustainability reports on their homepage, thereby making available a means to compare the reports and assess progress. Auchan and Kroger only supply the most current version online. This makes it difficult to assess developments in the environmental performance of the company as well as its degree of goal achievement with respect to sustainability. When the food corporations themselves compare their performance in the last year with previous years' reports, one must rely solely on this information. In other words, there is a lack of information accessibility and transparency.

To avoid these problems, Walmart has since 2008 published an annual progress report, in which it compares its GHG emissions from year to year. Interestingly, the emissions of Walmart and Tesco increased in 2009 and both of the retailers publish this development in their reports. This seems to be a positive sign for the transparency of information, albeit a negative one for their carbon footprint and the achievement of their goal to reduce CO_2. In contrast, the only information which can be found about the (weak) climate policy of Lidl, Aldi and Costco is on their homepage – and in all cases, it is difficult to date the information, or to know when commitments were made, as the retailers do not mention precisely when they started to commit to the environment.

In terms of targets, Tesco, Carrefour and the Metro Group have defined similar quantitative targets for reductions in GHG emissions. Walmart's "Sustainability 360" approach seems very ambitious; however, given the firm's increasing CO_2 emissions, they have yet to prove their ability to achieve this goal. Kroger has announced that it has already reduced its energy consumption by about 22 per cent and that they will achieve a 30 per cent reduction by 2010. In comparison to the other retailers, a large part of this reduction appears to have been achieved early and quickly. Here, further inquiries are needed to investigate the conditions which have allowed Kroger to make such progress in carbon emissions. The targets defined by Auchan, Aldi South, Rewe and Lidl are much less specific or apply to very specific sections of their carbon footprint, such as their electricity supply. As collective entities, the BRC and ERRT have defined quantitative targets as well. These targets apply to all members of the group jointly, so that they allow for over- and underperformers. In other words, the responsibility and accountability of the individual company for the given target is limited.

Most of the programs appearing to represent somewhat serious attempts to reduce GHG emissions on the part of a company or group are associated with investments in technologies and technological development, management and logistics as well as personnel. Tesco, in particular, has explicitly invested a large sum in a Sustainable Technologies Fund. In general, however, little detail on the financial implications and requirements of the programs can be found.

Many retailers cooperate with key actors in their institutional environment from civil society or the public realm in their efforts, thus adding legitimacy to their activities. This is not the case, of course, to the extent that such partners have the respective company as their sole customer, i.e. are entirely dependent on that company. Such a case exists with respect to the actor helping Rewe stores to obtain 100 per cent of their electricity supply from renewable sources. In addition, the cooperation with soft and broad initiatives like the Global Compact or the application of ISO 14000 standards promises little gain and therefore cannot be counted as contributing to the legitimacy and acclaim of a company's efforts. Finally, the cooperation with government can be evaluated quite critically, from a different perspective. Thus, the BRC's plans to inform the government on its climate-related activities, to work with national and international policymakers towards the establishment of a low carbon economy, and its support for global emission targets could, in the worst case, just reflect intensive lobbying activities, and possibly are aimed at lowering targets or prolonging timeframes envisioned by governments. Without further evidence on the actual content of the BRC's communication with and support of governments, such activities should not been seen as sources of legitimacy either.

As discussed in Chapter 3 by Hsueh and Prakash, information is crucial and, in our case, the investment of a number of retailers in carbon product labeling efforts is an interesting one for two reasons. First, such a measure would appear to allow a comprehensive and transparent approach to targeting GHG emissions associated with its products from farm to shelf. Such a measure transfers responsibility to the consumer, of course, which has both positive and negative sides. Second, the pattern in which the retailers considered here adopt this approach or invest in its development appears rather erratic. Tesco is quite advanced in this regard and has been publicly promoting its measures for a while now. Auchan, which generally does not appear to have a very ambitious climate-related program, is active in the area of carbon product labeling as well. Rewe has joined a partnership on the development of carbon product labeling with other food companies and think tanks. It is only involved with one product (strawberries) here, however, so that it still has to prove a real interest in such an approach. Interestingly, Carrefour has adopted a different approach. Instead of the carbon label, it uses the life-cycle approach so that a critical reflection of the two types of labeling is needed (Carrefour Group 2008). Through life-cycle assessment, Carrefour attempts to evaluate the environmental impacts of its products during the processes of production, use and disposal. This label includes, for example, raw materials such as oil or, in part, the emitted GHG. However, the carbon labeling which Tesco and Auchan have implemented indicates how much CO_2 is emitted from production to disposal. Tesco uses the standards of the Carbon Trust. The carbon labeling is also a life-cycle analysis, of course, but it does not measure the total environmental impact of aspects other than GHG emissions (Schmidt 2009). On the one hand, then, carbon labeling can be criticized in that it just evaluates the GHG emissions instead of all of the overall resource use associated with the production cycle of a product. On the other hand, however, it can be argued that GHG emissions are the main cause of climate change, which may be the most pressing problem we are facing, and that therefore it makes sense to create a label that prioritizes them.

However, in both cases the question whether the two types of assessment can reduce emissions of GHG still needs to be answered. The impact of the two labels has been widely criticized and declared to be inadequate (Schmidt 2009). In particular, the reduction of environmental impact through life-cycle analysis or the carbon label needs an adequately trained and informed consumer, as well as the provision of alternative consumption choices. Moreover, the calculations are always based on certain assumptions. The assessment of a carbon label for a can, for instance, is only applicable if consumers recycle the can. The two labels often calculate the GHG balance of recycled products. In addition, the calculation ends at the shelf, meaning that the distance which consumers drive to obtain a product is not included in the label. Therefore, we can see that labeling is still in the process of becoming an instrument to reduce the GHG emissions associated with food production and consumption.

On the basis of this comparison, then, there appear to be four rough categories of private programs of retailers and groups with respect to climate change issues. In the first group, retailers such as Walmart, Tesco and Carrefour find their place. They have ambitious goals, try many ways to reduce their environmental impact, and give good access to information and their climate policy achievements. The members of the second group also make information available, but it is not as detailed as that from the first group. Their activities and their commitments are weaker in comparison to the other retailers. Members of this group include Rewe, the Metro Group, Kroger, Auchan and the BRC (because of the lack of information and progress). The members of the third group do not make any clear commitments to climate change activities and do not regularly publish reports or make other detailed information about their climate activities available. Aldi South, Lidl (the Schwarz Group), Costco and the ERRT fall into this group. Finally, the members of the last and fourth group are those that neglect the issue of climate change altogether: Aldi North, the GlobalGAP, IFS and GFSI.

How do coverage and expected effectiveness of private retail food programs with respect to climate change match with the determinants of investments in private voluntary programs discussed in the theoretical framework laid out earlier in this chapter? In the following, we present some preliminary ideas on the importance of some factors versus others, keeping in mind, of course, that many of the programs reviewed are still only in infancy.

The visibility factor appears to play a role, but not a determinative one. All of the retailers considered are large. Yet, some of them do not even address their climate change impact in their private voluntary programs. Still, the larger retailers on average are more active. Their programs address the issue of climate change, even though the question of effectiveness cannot be answered with confidence for a large share of their activities. Additional explanations of the variance in the group derive from individual characteristics of the companies, which also influence their visibility. Walmart, for example, has been plagued by scandals and is in need of improving its image.

The characteristics of the home country also appear to play a role. Tesco has its home base in the UK, a country in which a lot of awareness and pressure on retailers with respect to their environmental and social performance exists. Likewise, the French chains, Carrefour and Auchan, are likely to be affected by the dialog between retailers and the government, mentioned above. In stark contrast, German retailers appear to experience the least pressure. In fact, the public perception of food discounters in Germany focuses almost entirely on questions of price and evokes hardly any discussion of environmental performance.[21] Even Lidl, which has experienced a number of labor scandals, apparently does not feel the need to really invest in private voluntary programs and redefine itself accordingly. Similarly, the characteristics of the host country appear to have an effect, as the German discounters, for instance, promote the small environmental efforts that they make only on their German homepages but not on the Austrian or Eastern European pages.

Finally, the involvement of public or civil society actors appears to aid the coverage and effectiveness of private voluntary programs with respect to climate change, even though the impact varies with the nature of the "partner". Neither partners depending solely on the partnering company for their existence (e.g. Rewe's partnership with German energy provider Energie-Handels-Gesellschaft), nor large networks with a lack of individual leadership and focus, such as the Global Compact, appear to be providers of a significant impetus. However, it is remarkable that the corporations cooperating with a public partner, specifically Carrefour, Walmart and Tesco, are relatively strong in their commitments. Here further research will be needed to investigate the influence of public institutions on the voluntary programs and standards of transnational corporations.

Additional and – in the context of our topic – particularly interesting influences appear to be provided by the characteristics of the problem of "climate change". Thus, it is noteworthy that the actors and groups which address climate change issues in their private voluntary programs the most are those with products in their range other than food. In other words, they may be experiencing public pressure because the public sees a link between these other products and energy consumption issues. The link between food production and consumption and climate change, in contrast, does not yet appear to be well established in the public debate.

This aspect is connected to the question of the costs involved in addressing a given problem. Not surprisingly, the majority of efforts delineated in the description of the private voluntary programs above focus on questions such as energy use in buildings, transport and travel. In the context of agrifood, these issues would appear to be the "easy" ones. Improvements in lighting or the efficiency of the truck fleet can easily be made with existing and affordable technologies. Even more extreme, the supply of customers with products needed to deal with changed weather conditions or providing information on private voluntary programs to the political realm offer opportunities for business profit and image campaigns rather than representing costs. In contrast, reducing overall transport needs is a tougher task in an agrifood system that is based on global sourcing and distribution chains. Reducing the climate contribution of food as such, for instance by reducing the share of meat sold and consumed, is even tougher, particularly as such an effort will not be popular with consumers. In this context, it is noteworthy that the cooperation between Tesco and the Carbon Trust in the development of product carbon footprinting does not include meat products.

What can we say about the determinants of group standards and their effectiveness? The ERRT and the BRC are groups of a medium size whose participants have a clear retail focus, thus providing some degree of homogeneity of interests, although clearly not perfect homogeneity. The interests of member firms, such as Ikea and Kingfisher, are quite likely to diverge on a number of issues. In comparison, the GlobalGAP, IFS and GFSI would seem narrower in their common focus, given that these are explicitly food-related programs. One may argue that these initiatives involve actors other than retailers, of course, thus broadening the diversity of interests again. Yet, critical analysis of the initiatives has shown them to be strongly dominated by retail interests, especially during the stage of the schemes' creation.

Given the existing, although limited, heterogeneity of participating members in the BRC and ERRT, our theoretical considerations would lead us to expect a standard in the mid-range, allowing for over- and underperformances. This situation is indeed reflected in the lack of sanctioning and enforcement provisions of the two private voluntary programs and their provisions with respect to climate change, as well as the definition of rather broad objectives for the groups as such, rather than specific aims for their individual members.

Again, it is noteworthy that the private voluntary group programs focusing explicitly on food are those which are the least active with respect to climate change. This appears to be another indication that the link between food production and consumption and climate change is not yet well established in the public debate and consumer awareness. It remains to be seen whether this situation will change with the increasing focus on the agrifood contribution to GHG emissions which we are currently witnessing. In other words, our expectation that a higher degree of homogeneity of the group and therefore less collective action problems would induce a private voluntary program, in which the standards are more narrowly defined, has to be expanded to include the interests of the given group members. If the dominant preference of the group members is not to invest in a respective program, then the greater homogeneity of the group will only make a weaker output more likely.

Conclusion

This chapter has shown that the climate change related commitments and activities by major retailers or relevant groups and their private voluntary programs are diverse. They range from ambitious quantitative targets to a complete neglect of climate change issues. Importantly, only a few programs mention compliance and monitoring processes in terms of their own rules. Because of the lack of control mechanisms, then, even the schemes promising to be ambitious are not able to engender sufficient trust that a "real" change towards a reduction in GHGs will materialize.

How do these findings relate to the program performance, which the theoretical framework laid out in this chapter would have us expect? Visibility appears to be a factor, as the biggest retailers are also particularly active. It is clearly not a determinative one, however, as those retailers not engaging in private voluntary programs related to climate change are far from small. The home country effect appears to be quite powerful, as does the influence of (serious) partners. Overall, then, a range of factors appear to induce retail corporations to invest in private voluntary programs relating to climate change. However, they do so to varying degrees and, for the majority of retail corporations, to a limited extent only. Particularly relevant, moreover, appear to be the problem characteristics and their influence on the costs of potential remedies. In the case of climate change, these problem characteristics mean that measures targeting the major sources of GHG emissions in the agrifood sector are extremely costly.

From the perspective of the framework, we are facing a situation in which relatively steep cost curves are combined with benefit curves which, for many retailers, appear to be set at relatively low levels. The conclusions which we can draw from this analysis are twofold. We can identify a potential contribution of private voluntary programs in the retail sector to the pursuit of climate change objectives, in general. We also have to acknowledge, however, that this contribution is very limited right now and unlikely to expand dramatically in the future, unless conditions change substantially.

To be more specific, private voluntary programs can contribute in certain ways and to a certain extent to climate change governance in the agrifood system. With sufficiently specified targets, a sufficient degree of transparency and monitoring as well as the specification of sanctioning mechanisms, private voluntary programs can probably contribute to increasing the energy efficiency of buildings, storage and transport, for example. In other words, such programs can foster the diffusion of available and affordable technologies, as well as corresponding organizational and logistical measures. In order to foster the adequate design of private voluntary programs across the board, however, an appropriate public framework is probably needed.

Moreover, the link between food production and consumption and climate change needs to be better established in the public debate. In consequence, the investigation and highlighting of this link should probably be a prime target for political and scholarly intervention in order to foster climate change governance in the agrifood system. With increasing public awareness of the large impact of food production and consumption, consumers and citizens would be able to exert more pressure on food retail corporations to pursue serious efforts in this respect. Similarly, public actors may consider establishing a firmer link between food policy, consumer policy and climate policy. Currently, these policy types frequently are handled in different ministries or by different public agencies. A better integration would not only allow a more systematic targeting of the GHG emissions associated with food production and consumption by public governance, which in itself is an urgent necessity. It would also raise awareness of this link by consumers as well as business actors connected to the food chain.

Yet, private voluntary programs should not be expected to address and solve such fundamental problems as the rise in GHG emissions caused by the increasing role meat plays in the diets of Western consumers as well as wealthy consumers in developing countries, for instance. Even the extent to which private voluntary programs will be able to really reduce food miles in a global agrifood system based on global sourcing and distribution structures is questionable. To make necessary fundamental structural changes, public actors will have to take on the responsibility themselves. This is particularly the case as consumers face enormous information and collective action problems, of course, when attempting to change the carbon footprint of the global agrifood system from the consumption side.

If the necessary structural changes in the global agrifood system are in fundamental opposition to the interests of today's powerful retail corporations, one may argue that looking at their private voluntary programs to assess such activities in the food sector is the wrong approach. Maybe we should look at the activities of other actors in the production and value chain instead. Indeed, individual food processors such as Nestle and Kraft have made commitments towards climate change. Similarly, individual pig farmers in Germany are experimenting with combining pig raising with biogas production. The power of retail corporations and the private voluntary programs which they have set up are one of the few available systems for a systematic implementation of climate change activities in the global food system so far, however.

Notes

1. These numbers include so-called CO_2 equivalents. These equivalents describe the consequences of other gases, which are far more dangerous for the atmosphere than CO_2, such as methane or nitric oxide, for climate change. They are calculated in relation to CO_2. Without the calculation of CO_2 equivalents the climate change impact of the agrifood sector would be less than 20 per cent (Greenpeace 2008).
2. Private voluntary programs with rules and norms for business may also be initiated by civil society actors, of course.
3. Private voluntary programs covering the interaction between the agrifood system and climate change, in the broadest sense, are self-regulatory schemes in the field of biofuel production. However, these programs represent a special case and do not reflect the broader dynamics of this interaction. Accordingly, they will not be considered here.
4. While private voluntary programs, in general, have played a major role in the global agrifood system for a number of years, climate change related programs by retailers are new arrivals on the scene.
5. Depending on the exact interpretation, the term "voluntary" may be misleading, at least in some cases (see below).
6. As the costs of implementation and documentation (including auditing and certification) tend to be extremely high from the perspective of small farmers in developing countries, private retail food standards have received a lot of criticism for pushing these farmers out of the market and into subsistence farming.
7. The list of factors also shows that socio-institutional path dependencies will exert an influence.
8. With increasing concerns about the capture of civil society actors, their legitimacy claims may also be challenged, of course.
9. Such calculations must be seen in a mid- to long-term perspective, of course, because some investments are only likely to pay off after a while. This will most certainly be the case for the actual achievement of climate change objectives. Benefits drawn from a better image of a retail corporation or the retail sector as such may well accrue (or have accrued) in the short run.
10. Technical support for company B by company A is also possible in this context.
11. Aldi is split into Aldi North and Aldi South, which between them cover territories in Germany and internationally. Aldi North operates in Belgium, Denmark, the north of Germany, France, Luxembourg, Portugal, Poland and Spain. Aldi South operates in Australia, Austria, the south of Germany, Greece, Hungary, Ireland, Slovenia, Switzerland, the United Kingdom and the US.
12. In this chapter, we consider the commitments of Lidl, as the biggest food retailer in the Schwarz Group.
13. Although Kroger operates in the US only and is not a transnational supermarket chain, we mention it here because of its economic power and number of existing stores. It is assumed that the environmental impact of Kroger is comparable with the other retail food corporations.
14. The authors are aware that some of these international standards cannot ensure program stringency, as they entail weak control or sanctioning mechanisms, as discussed in other chapters in this book. However, this criterion was chosen as such standards allow a first step towards a comparison and benchmarking of programs.
15. The Greenhouse Gas Protocol was established by the World Resources Institute and the World Business Council for Sustainable Development and aims to foster the development and promotion of internationally accepted accounting and reporting systems for greenhouse gases (Anon 2010; ⟨http://www.ghgprotocol.org/⟩).

16. Through carbon labeling it is possible to evaluate how much CO_2 and its equivalents is emitted during the production, transportation, use and waste process of selected products (Tesco 2009). Consumers are informed about the GHG which were produced when they buy carbon-labeled products, so that they can play an active role in decreasing emissions (Tesco 2009).
17. A criticism may be leveled that Tesco monitors its own activities in this respect. The fact that Tesco indicates that it emitted more CO_2 in 2009 than in 2008 is a hint that the monitoring process is independent.
18. The relevant members who participated in this commitment are Asda (Walmart in the UK), the Carrefour Group, the Metro Group and Tesco.
19. The Metro Group has published a climate brochure which, however, provides much less information than the other reports.
20. It is also possible to order all the reports (except for the most recent one).
21. A frequently voiced view is that the discounters provide cheap, quality products. Quality, however, does not refer to environmental or social characteristics in this context.

REFERENCES

Aldi South. 2009a. Energie gewinnen. ⟨http://verantwortung.aldi-ued.de/verantwortung/html/energie_gewinnen.htm⟩; accessed June 15, 2009.

Aldi South. 2009b. Energie Sparen. ⟨http://verantwortung.aldi-sued.de/verantwortung/html/energie_sparen.htm⟩; accessed June 15, 2009.

Aldi South. 2009c. Klima schonen. ⟨http://verantwortung.aldi-sued.de/verantwortung/html/klima_schonen.htm⟩; accessed June 15, 2009.

Anon. 2010. "What is the Greenhouse Gas Protocol?" ⟨http://www.ghgprotocol.org/about-ghgp⟩; accessed February 10, 2010.

Auchan. 2009. Entreprise Responsible 2008/2009. Paris. ⟨http://www.auchan.fr/univers/articles/universArt.jsp?idChannelN1=21950&docId=42452⟩; accessed June 15, 2009.

Biedermann, R. 2007. *Sozialstandards durch Private Governance*. Baden-Baden: Nomos.

Bonanno, A., L. Busch, W. Friedland, L. Gouveia, and E. Mingioni. eds. 1994. *From Columbus to ConAgra: The Globalization of Agriculture and Food.* Lawrence KS: University Press of Kansas.

British Retail Consortium. *A Better Retailing Climate.* ⟨http://www.brc.org.uk/showDoc04.asp?id=3558&moid=5603⟩; accessed June 15, 2009.

Burch, D., and G. Lawrence. 2005. "Supermarket Own Brands, Supply Chains and the Transformation of the Agri-Food System." *International Journal of Sociology of Agriculture and Food* 13 (1): 1–18.

Carrefour Group. 2005. *Sustainability Report 2004: Promoting Socially Responsible Commerce and Constructive Globalization.* Paris. ⟨http://www.carrefour.com/docroot/groupe/C4com/Commerce%20responsable/Publications/RDD2004uk.pdf⟩; accessed June 15, 2009.

Carrefour Group. 2008. *Sustainability Report 2007.* ⟨http://www.carrefour.com/docroot/groupe/C4com/Commerce%20responsable/Publications/RDD%202007%20GB.pdf⟩; accessed May 08, 2009.

Carrefour Group. 2009. *At the Heart of Life.* ⟨http://www.carrefour.com/docroot/groupe/C4com/Commerce%20responsable/Publications/RGG2008GB.pdf⟩; accessed June 15, 2009.

Carrefour Group. *Limiting the Environmental Impact of Our Stores.* ⟨http://www.carrefour.com/cdc/responsible-commerce/our-commitment-to-the-environment/reducing-stores-impact/?com.carrefour.cdc.print.page.content=true⟩; accessed May 8, 2009.

Clapp, J., and D. A. Fuchs. eds. 2009. *Corporate Power in Global Agrifood Governance*. Cambridge MA: MIT Press.

Cutler, A. C., V. Haufler, and T. Porter. eds. 1999. *Private Authority and International Affairs*. Albany: State University of New York Press.

Dixon, John. 2007. "Supermarkets as New Food Authorities." In *Supermarkets and Agri-food Supply Chains: Transformations in the Production and Consumption of Foods*, eds. David Burch and Geoffrey Lawrence, 29–50. Cheltenham, UK: Edward Elgar.

Drache, D. ed. 2001. *The Market or the Public Domain?* London: Routledge.

Enquete Commission Globalization of the World Economy. Deutscher Bundestag. 2002. *Schlussbericht der Enquete-Kommission Globalisierung der Weltwirtschaft*. Opladen: Leske & Budrich.

Fuchs, D. 2003. *An Institutional Basis for Environmental Stewardship*. Dordrecht: Kluwer Academic Publishers.

Fuchs, D., A. Kalfagianni, and M. Arentsen. 2009. "Retail Power, Private Standards and Sustainability in the Global Food System." In *Corporate Power in Global Agrifood Governance*, eds. J. Clapp, and D. A. Fuchs, 29–59. Cambridge MA: MIT Press.

Gibson, R. ed. 1999. *Voluntary Initiatives*. Peterborough: Broadview Press.

GlobalGap. 2009. *GlobalGap and SQF to work in partnership, SQF to benchmark SQF 1000 code with GlobalGap*. ⟨http://www.globalgap.org/cms/front_content.php?idcat=44&idart=756⟩; accessed December 2, 2011.

Graz, J. C., and A. Noelke. eds. 2007. *Transnational Private Governance and Its Limits*. London: Routledge.

Greenpeace. 2008. *Cool Farming: Climate Impacts of Agriculture and Mitigation Potential*. Amsterdam. ⟨http://www.greenpeace.at/uploads/media/Cool_Farming_Report_Final_web.pdf⟩; accessed December 10, 2009.

Konefal, J., M. Mascarenhas, and M. Hatanaka. 2005. "Governance in the Global Agro-Food System: Backlighting the Role of Transnational Supermarket Chains." *Agriculture and Human Values* 22 (3): 291–302.

Kroger. 2008. Doing our Part. 2008 Sustainability Report. ⟨http://www.thekrogerco.com/documents/KrogerSustainReport08.pdf⟩; accessed May 14, 2009.

Lidl. Umwelt/Nachaltigkeit. ⟨http://www.lidl.de/cps/rde/xchg/lidl_de/hs.xsl/68535.htm⟩; accessed June 15, 2009.

Metro Group. 2008. *Climate Action: Carbon Footprint Strategy.* Dusseldorf. ⟨http://www.metrogroup.de/servlet/PB/show/1172910_ainline/Verantw-Nachh-Umwelt-Klimaschutzbroschuere-en.pdf⟩; accessed June 15, 2009.

Nash, J., and J. Ehrenfeld. 1997. "Codes of Environmental Management Practice." *Annual Review of Energy and Environment* 22(1): 487–535.

Porter, T. and K. Ronit. eds. 2010. *The Challenges of Global Business Authority: Democratic Renewal, Stalement or Decay?* New York: State University of New York Press.

REWE. Vorreiter bei Grünstrom. REWE deckt annähernd ihren gesamten Energiebedarf aus erneuerbaren Energien. ⟨http://www.rewe.de/index.php?id=4585⟩; accessed June 15, 2009.

Schmidt, Hans-Jürgen. 2009. "Carbon Footprinting, Labelling and Life Cycle Assessment." *International Journal of Life Cycle Assessment* 14 (1): 6–9.

Tesco. 2009. *Corporate Sustainability 2009*. Cheshunt, UK: Tesco. ⟨http://www.investis.com/plc/cr09/crr09.pdf⟩; accessed June 15, 2009.

Vogel, D. 2005. *The Market for Virtue: The Potential and Limits of Corporate Social Responsibility*. Washington DC: Brookings Institutions.

Walmart. "Wal-Mart & Clinton Climate Initiative Announce Partnership." ⟨http://walmartstores.com/Sustainability/8141.aspx⟩; accessed January 7, 2010.

Walmart. 2008a. *The Company of the Future*. Fact Sheet. ⟨http://walmartstores.com/download/3063.pdf⟩; accessed May 7, 2009.

Walmart. 2008b. "Wal-Mart Promotes Sustainable Agriculture." ⟨http://walmartstores.com/download/2307.pdf⟩; accessed May 7, 2009.

Walmart. 2009. *Global Sustainability Report: Now More than Ever*. ⟨http://walmartstores.com/Sustainability/7951.aspx⟩; accessed May 7, 2009.

6

Climate change, private voluntary programs and the automobile industry

Tony Porter

Introduction

Cars and trucks are estimated to contribute to between 15 and 20 per cent of worldwide CO_2 emissions (IEA 2009: 3, 29). With the rapid growth of automobile ownership in the developing world this is likely to rise. Thus controlling auto emissions is a crucial part of addressing the problem of climate change. However, reducing auto emissions poses unique challenges. This includes technical challenges associated with the leap from the conventional internal combustion engine to new types of engines and energy sources. It also includes unique organizational and policy challenges, since solutions must involve individual consumers operating mobile sources of emissions; an auto industry emblematic of "Fordist" high-wage, high-quality industrial employment with dispersed transnational supply chains that is especially vulnerable in the current economic crisis; a technology that has complicated systemic linkages to other policy fields and industries, including oil, urban and infrastructure planning and management, public transportation; and the car's powerful cultural grip on feelings of individual mobility, power and identity. All of these operate globally, but in quite different ways in different locations. In other words, when thinking about climate change we need to think of the automobile as about far more than the vehicles themselves or even the industry that manufactures them, but instead to see it as dependent on a larger ensemble of practices, relationships, objects, infrastructures, energy sources, engineering knowledge, cultural attitudes and supply

Business and climate policy: The potentials and pitfalls of private voluntary programs,
Ronit (ed.), United Nations University Press, 2012, ISBN 978-92-808-1214-5

chains, all of which work together as a system that has been usefully labeled *automobility* (Featherstone, Thrift and Urry 2005; Paterson 2007). The challenge of climate change requires a change not just in the vehicles, but in this broader system of automobility.

Considering the complexity of the system of automobility and the challenges involved in changing it, it is clear that the active engagement of business actors is essential if the industry's damaging impact on the climate is to be reduced or reversed. Even if compulsory public regulations are strengthened dramatically it would still be impossible for them to specify in any detail the enormously varied and interdependent changes that would be needed. As discussed in the concluding chapter, private voluntary programs can usefully be treated as displaying great variation in the degree to which they are private, voluntary and institutionalized as programs. In the case of the automobile it is important to understand the diversity of private voluntary programs that exist and the parts that they can be expected to play or not play in engaging business actors to address climate change. For example, in order to succeed, a government-initiated and government-funded program may require extensive engagement, self-organization and initiative on the part of business and we may treat this as a voluntary program with a crucial degree of privateness, which is quite different from traditional compulsory state regulation of business.

A great many academic literatures and official reports have moved beyond simplistic state/market distinctions to explore the tremendous variation that is evident in the character of public governance and regulation, in the ways that business is organized and contributes to governance, and in the complex relationships that have developed between public and private authorities. In an increasing number of issue areas, including climate change, the state will fail to mobilize the necessary response if it acts too independently or peremptorily and if market actors act too independently they will fail to achieve the necessary cohesion and seriousness. In addressing the problem of automobility and climate change it is important to think about the lessons that these broader literatures offer and the lessons that efforts to address this problem can offer to these broader literatures. A key insight from existing literatures is that there is a strong relationship between the character of the problem to be solved and the character of the rules and institutions that are needed to align incentives so that the relevant actors will be motivated to solve it. This highlights the importance of building on the capacities of existing institutions, matching them as closely as possible to the types of tasks to which they are best suited.

First, generally we may predict that the state's unique capacity to manage binding commitments across multiple settings and long timeframes involving large numbers of actors will be necessary for the aspects of problems for which private and other and often smaller institutions are too weak or lack organizational capacity, although this may not be so for some small, weak states. Second, we may predict that private voluntary programs will be especially important for addressing more focused, detailed problems in rapidly changing environments. A third prediction that can be drawn from existing literatures is that when shifts in policy paradigms are needed, neither the state nor business by themselves can bring about this change, and broader public involvement in norm change is needed (Finnemore and Sikkink 1998; Hall 1993). In this chapter we will see that these predictions are confirmed in the successes and failures of efforts that have been made so far to address the problem of automobility and climate change.

In this process of matching institutions to tasks the full range of public, public–private and private institutions will likely need to be mobilized, with new initiatives building on existing institutions. This is especially the case with a complex system such as automobility. For any particular institution to successfully contribute to solving climate-change related collective action problems, it is necessary for the four functional stages of programs set out in Chapter 1 – designing, joining, monitoring and complying – to be carried out. The challenge of transforming automobility involves so many interdependent tasks, and such a wide mix of public and private aspects, that in many cases these four functional stages will not be stand-alone properties of a single voluntary program but instead will draw on aspects of the larger system that facilitate the provision of that function for that program. For instance, government threats to step in if voluntary programs fail can bolster compliance. Similarly the refusal of assemblers to accept non-compliant products from their suppliers can motivate compliance with voluntary initiatives of suppliers.

The next section provides an overview of public programs that have been developed to address automobility and climate change, and the private voluntary aspects that are important within them. The chapter then examines more purely private voluntary programs. The final section concludes by returning to the role of the functions of designing, joining, monitoring and complying in private voluntary programs, as well as drawing other lessons from the cases, especially with regard to the three predictions set out above.

The development of public programs towards automobility and climate change, and private voluntary aspects within those

In order to understand the role of private voluntary programs it is necessary to start by reviewing the public-initiated programs that have been developed, with their important private voluntary aspects. In this section

I examine these public programs with voluntary private aspects, looking first at the US, then at the EU, and finally at other jurisdictions. While all jurisdictions have a role to play in addressing the relationship between automobility and climate change, the US and the EU, because of their political and economic importance in the global political economy, are especially crucial, and thus particular attention is devoted to their programs. Following the discussion of national programs, this section concludes by discussing transnational programs initiated by public authorities with private voluntary aspects.

Concerns about auto emissions date back to the 1970s, although at that time the focus was on smog and acid rain, along with lead. In the US concerns about these combined with worries about dependence on foreign oil after the OPEC cartel dramatically raised prices in 1973. Since there are limits to the degree to which gasoline can be re-engineered to eliminate unwanted pollutants, there is a close correlation between the reduction of these pollutants and fuel economy. The US Energy Policy and Conservation Act of 1975 created the Corporate Average Fuel Economy (CAFE) mandatory standard. Japan followed with fuel economy rules later in the 1970s. Serious attention to emissions standards in Europe only began in the late 1980s, stimulated especially by concern at the effect of acid rain on forests (Shiroyama 2007). Since those early days, concerns about vehicle emissions and the efforts of public authorities to control them have grown, but in ways that differ significantly across jurisdictions. Three main overlapping sets of policy responses to the problem of vehicle emissions can be identified.

The first set includes the efforts to restrict emissions at the tailpipe, requiring incremental changes in existing automotive technologies. These policy responses seek to intervene at the output end of the production process, which can then stimulate technological changes throughout the production chain.

The second set includes the efforts to make the transition to entirely new, greener technologies. This set most directly intervenes at the beginning of existing and new production processes. The most immediately feasible new technological solutions are: advanced diesel, which the European manufacturers have pioneered and which has been widely adopted in Europe; hybrid gasoline/electric engines, which the Japanese manufacturers have pioneered and which are slowly making inroads with consumers despite their significantly greater cost than conventional engines; and biofuels, which have been especially popular in the US. Other more ambitious technologies include hydrogen fuel cells and fully electric cars.

The third set of policies aims to change the use of vehicles by intervening in areas other than the process of producing vehicles. These include, for instance: intelligent transport systems that organize traffic more efficiently and reduce congestion; initiatives to promote greener alternatives to automobility, such as trains, bicycles or walking; the introduction of road tolls; and reworking urban planning models. This set of policies does not focus on automobiles directly and will only be mentioned in passing.

US public policies with private voluntary aspects

In this section we shall see that the most important US initiatives had significant, but relatively weak, private voluntary components. There were serious shortcomings in the government initiatives, but I shall argue later on that the weakness of the private voluntary components also contributed to the poor climate-related performance of the US automobile industry. The unevenness and lack of integration of private voluntary programs, which emerges as a major theme in this book, begins to be evident in this section.

In the United States, regulation of tailpipe emissions has primarily focused on the mandatory Corporate Average Fuel Economy (CAFE) standards. CAFE is the average fuel economy of a manufacturer's production in a model year (its "fleet"), calculated by weighting each model's emissions by its sales. The voluntary private aspect of the CAFE program is the great discretion that manufacturers are allowed in how they comply with the compulsory standard. When introduced in 1975, the goal of the CAFE standards was to double new car fuel economy to 27.5 miles per gallon by 1985. The standards for light trucks, which came to include sport utility vehicles (SUVs), were set much lower. In its first decade the CAFE standards enjoyed some success: the average fuel economy of passenger cars and of light trucks each improved more than 60 per cent (Onoda 2008: 28). Unfortunately, the average of both together declined significantly over time because of a massive shift of car buyers from passenger cars to SUVs and other light trucks (An and Sauer 2004: 9), and because the CAFE standards were not raised to offset this shift. Aggressive lobbying by the industry, the competitive benefit conferred on US producers by their shift into light trucks, and skepticism towards climate change, especially during the Bush Administration from 2000 to 2008, contributed to the failure of the US industry and government to restrain emissions.

When Obama took office he replaced the head of the Environmental Protection Agency (EPA) with Lisa P. Jackson, more favorably inclined towards regulating emissions, and asked the agency to reconsider its denial under Bush of California's request to regulate them. The bankruptcy of GM and Chrysler in the economic crisis dramatically strengthened his hand, through the financial control over those firms that it brought the federal government and by demolishing the credibility of the auto executives. Then in May 2009 Obama, with the auto executives and environmentalists at his side, announced a plan whereby greatly strengthened CAFE standards would be introduced in 2011. Many environmentalists and others lauded this as a historic negotiating accomplishment. According to the Sierra Club it was "one of the most significant efforts undertaken by any president, ever, to end our addiction to oil and seriously slash our global warming emissions" (*The Australian* 2009). Others thought the automakers looked like "prisoners of war being paraded before the press by their captors" (Winter 2009), although the industry certainly benefited from having a single national standard rather than a patchwork. According to one administration official, the new rules would reduce carbon dioxide emissions by 900 million tonnes over five years, the equivalent of taking 177 million cars off the road or shutting down 194 coal plants (LoBianco 2009).

Public authorities in the US have also pursued the second set of policies identified above, intervening to promote new, greener or more energy-efficient technologies. This has primarily been pursued through public–private programs that are initiated by the government but rely heavily on the active participation of private actors, thereby giving these programs a private voluntary aspect. The Partnership for the Next Generation of Vehicles (PNGV), introduced by the Clinton Administration in 1995, was the first national public–private program. The PNGV aimed to develop new technologies that would reduce emissions, increase fuel efficiency and create an affordable line of vehicles that balance efficiency with comfort and reliability. It set a target of developing vehicles with 80 miles per gallon fuel efficiency by 2004 (Malakoff and Service 2002). In 2001, the Bush Administration replaced the PNGV with FreedomCAR (Cooperative Automotive Research) and shifted the focus of the program more exclusively to the development of hydrogen fuel cells. The more specific targets of the PNGV were replaced with longer-term and less specific goals. The original PNGV partners were the US Department of Energy and the US Council for Automotive Research (USCAR).[1] Also in 2003 President Bush announced a complementary Hydrogen Fuel Initiative for which he would request $720 million in funding. In 2007 plug-in hybrid electric vehicle (PHEV) technologies were added as a FreedomCar priority, and this has continued with efforts to support President Obama's goal of 1 million plug-in hybrid vehicles by 2015.

The US Department of Energy's Vehicle Technologies (VT) program, which sponsors FreedomCar, also sponsors other research, such as alternative fuels other than hydrogen (US DOE 2009a). It sponsors the Clean Cities program, established in 1993, which brings together roughly 90 coalitions, government agencies and private companies in pursuit of reducing oil imports, regional economic development, and better air quality. It claims to have reduced petroleum consumption by 2.4 billion gallons from 1993 to 2009. It has a particular interest in the infrastructure for alternative fuels. The 21st Century Truck Partnership, a similar program to FreedomCar but for heavier trucks and buses, was created in 2000. California has also been especially active in building public–private programs promoting hydrogen technology. For instance, the California Hydrogen Highway Network (CaH2Net), initiated in 2004, is a public-sponsored program with private participation directed by the California Air Resources Board, which has aimed to make hydrogen fueling available along California's major highways and to have as much as possible of that hydrogen produced from clean, renewable energy (State of California 2009).

From the beginning, FreedomCar was controversial. Environmentalist groups saw the initiative as a way for the auto industry to avoid strengthened regulation of emissions. Instead of forcing automakers to reduce emissions, the US government was financing a highly speculative technology that might provide valuable future commercial benefits to automakers, but would involve no mandatory adjustment in their present or future production. The enormous costs involved in hydrogen research could instead have been used for environmental solutions with more immediate and certain payoffs. Clean energy that might be used to produce hydrogen could be used to displace carbon-intensive sources of electricity with more positive effect on the emission of greenhouse gases (GHGs) (Hammerschlag and Mazza 2005). The program was also criticized for the large number of earmarks devoted to specific political interests that were not central to the program's goals, estimated to be 40 per cent or more of the budget (Rohde 2005; US GAO 2008: 6; Yacobucci 2008). In its major second review of the FreedomCar and Fuel Partnership program in 2008, the National Research Council (NRC), while noting some progress in some areas, made clear that the program's goal of having a commercially viable hydrogen vehicle on the road by 2015 was far from certain: "In summary, progress has been good in most areas and impressive in a few. However, resolving the barriers already known as well as those yet to be uncovered will clearly present major challenges" (National Research Council 2008: 110). In his budget for 2010, President Obama included no funding requests for hydrogen research, an indication of a shift in emphasis to electric plug-in vehicles and biofuels and a loss of enthusiasm for hydrogen (*Oil & Gas Journal* 2009: 18).

The US government has also provided a lot of support for biofuels. The private voluntary aspect of these programs is evident in their reliance on private firms and associations to develop the biofuel industry. Business interest associations include the Renewable Fuels Association (RFA), the

American Coalition for Ethanol (ACE), the National Biodiesel Board (NBB) and the National Corn Growers Association (NCGA). These associations engage in lobbying, but also work to develop the industry. For instance, the RFA provides technical data, promotes continuous improvement in the industry, and helps set standards, and the ACE provides technical information about storage and fueling systems. Significant growth in the share of ethanol in the US gasoline supply only began after the turn of the millennium, increasing from 1 per cent in 2000 to 7 per cent in 2008 (Westcott 2009: 29). In 2006 the US overtook Brazil to become the world's largest producer of ethanol (Pryor 2009: 523). This growth was driven by high energy prices and government subsidies. The Energy Independence and Security Act of 2007 provided especially strong support for biofuels, setting a 2022 target of 36 billion gallons, five times the target set out in the Energy Policy Act of 2005 (Westcott 2009: 30). However, the biofuel industry has faced serious challenges and criticisms. It was widely seen as having contributed to the rise in food prices in 2007, and the consumption by biofuels of 24 per cent of the US corn industry in 2007/8 contributed to this perception (Westcott 2009: 29). In 2009, preliminary studies by the EPA suggested that corn ethanol produced more GHGs over its life cycle than gasoline, raising questions for its eligibility for consideration under the 2007 Act (*New York Times* 2009: 7). The majority of new ethanol plants in the US have been designed to use coal to produce the ethanol (Akinci, Kassebaum, Fitch and Thompson 2009: 3488). The substitution of biomass that does not compete with food production faces serious technical challenges, as does the development of fueling infrastructure (Westcott 2009).

As we shall see below with respect to other regions, the complex changes required in the US system of automobility will only be possible with effective interaction between strong public policies and strong industry engagement in private voluntary programs. If the government simply tries to threaten or bribe the industry to meet particular simple government targets, it will get shallow compliance at best. This has been the experience in the US.

EU public programs with private voluntary aspects

Since the 1990s the EU's efforts on vehicle emissions have been driven by the conflicting pressures it has experienced in meeting its Kyoto commitments, and resistance led by the industry's aggressive lobbying that has claimed that meeting these commitments would do unacceptable damage to the industry's competitiveness and employment levels. In the late 1990s, faced with no forward movement on fuel efficiency since the mid-1980s, the European Commission negotiated voluntary agreements with manufacturers, which are examined in the next section below. By 2007 it had become clear that the voluntary program was not achieving reductions at the pace needed if the EU was to meet its international commitments on climate change. Although average emissions from new cars had dropped by 12.4 per cent between 1995 and 2004, at 163 g CO_2/km it was clear that neither the 140 g target by 2008/9 or the 120 g target by 2012 would be reached (Onoda 2008: 25). EU GHG emissions overall had declined by 5 per cent over the 1990 to 2004 period while CO_2 emissions from road transport had increased by 26 per cent (European Commission 2007: 2). Critics argued that the industry had done little better than "business as usual". Part of the problem was also that the technical gains in efficiency had been offset by the use of these gains to produce and sell more-powerful cars.

In response to these problems, the EU established a mandatory target of 120 g CO_2/km by 2012. The automakers were expected to achieve 130 g CO_2/km with the remaining 10 g CO_2/km coming from other measures such as tire improvements or the use of biofuels. Debate over these targets and their implementation was intense. Environmentalists saw the inclusion of 10 g CO_2/km from other improvements as a capitulation to the industry's lobbying for a "comprehensive" approach that would shift the burden to other actors. They also criticized the linking of targets to vehicle weight, which gives heavier vehicles easier targets relative to other measures. The industry issued dire warnings about the impact of the standards on competitiveness and employment. For instance, DaimlerChrysler threatened the loss of 65,000 jobs if the standards were implemented (Corporate Europe Observatory 2007). The determination of governments to defend and promote the interests of firms headquartered in their jurisdictions provoked political conflicts, especially between Germany, which has specialized in more powerful, higher emission cars, such as BMW and Daimler, and France and Italy, which have specialized more in low-emission cars, like Peugeot/Citroen, Fiat and Renault (ICCT 2007: 14).

Quite importantly, and within this compulsory approach, there remained some private voluntary aspects in how it would be implemented. For instance, the industry was permitted to pool its targets so, for instance, high-emissions BMW could work with low-emissions Volkswagen to produce an acceptable average level of emissions. Although there have been proposals to link this regulatory system to the EU's larger interest in emissions trading schemes (Michaelis and Zerle 2006) this has not yet occurred.

The EU has a long history of promoting research and development collaboration among leading EU firms, such as the EUREKA framework for R&D collaboration, launched in 1985, which covered a wide variety of industries, including transport. As in the US case, these public programs are designed by public authorities with private input, but rely upon active private participation in their implementation, and this gives them a private voluntary aspect. In general these programs use positive incentives, such as government financing of R&D and the provision by public authorities of opportunities for collaboration that are valuable to the industry. Given the uncertainty associated with the technologies, targets are often not specified in great detail, and the need to change targets as research progresses is recognized. Accordingly, the monitoring and compliance elements that are important in other programs do not display the same importance here.

Although the primary motivation has often been to foster competitiveness, some of the EU R&D programs have supported new, greener technologies, such as with ERTICO, created in 1991, a joint venture between the European Commission, ministries of transport and industry participants that researches intelligent transport systems (ERTICO 2005). The European Platform for Cooperation and Coordination in Transport Research (EPTR), created in 2001, consists of government representatives, but it funds the networking costs of its non-governmental partners. In 2002, following the US launch of FreedomCar, the European Commission announced a plan to spend €2.12 billion from 2003 to 2006 on renewable energy, with a focus on hydrogen (Miller, Bahree and Ball 2002: A16). ERA-NET Transport, created in 2004, brings together actors involved in transportation research, in the context of the larger ERA-NET effort to coordinate national research programs (ERA-NET Transport 2005).

In 2008 the EU launched a public–private joint technology initiative on hydrogen fuel cells with projected funding of €470 million from the EU's Seventh Framework Program and with a matching amount to be provided from private sources. In 2009 the European Commission launched its European Green Cars Initiative (EGCI) as one of the three public–private programs included in its recovery package. The funding that this initiative would mobilize was anticipated to be €5 billion, mostly in the form of loans from the European Investment Bank (EU 2009).

One of the most important institutions promoting collaboration on R&D in the automotive industry is EUCAR, created in 1994 by the major auto manufacturers. EUCAR involves other stakeholders, including public sector authorities, but it is an industrial association owned by its members, which are the 12 major European manufacturers of cars, trucks and buses. It has been very active in a wide variety of research relevant to climate change, including hydrogen systems. In 2009, together with CLEPA, the European Association of Automotive Suppliers, it proposed a set of priorities for the EU's European Green Car Initiative (EUCAR and CLEPA 2009).

The EU has provided a large and increasing amount of support for biofuels to meet its climate change commitments, to reduce dependence on oil imports, and to improve the welfare of its agricultural sector. As in the US, this public program relies heavily on private actors to develop the industry, giving it a private voluntary aspect. For instance the industry-led European Biofuels Technology Platform (EBTP), established in 2006, is a public–private program that works to develop the industry. Other relevant industry associations include the European Biodiesel Board (EBB) and the European Bioethanol Fuel Association (eBIO). The major focus of these associations is lobbying, although the EBB's Biodiesel Quality Report, which requires sampling and independent testing from all members, seeks to ensure the quality of the industry's product.

EU biofuel production, mostly biodiesel, increased from just under 1 million tonnes in 2000 to 5.9 million tonnes in 2006, but even then it only provided 1.8 per cent of the EU's transportation fuel (Kutas, Lindberg and Steenblik 2007: 5). Since the sustainability of automobility technologies such as biofuels, hydrogen or electric cars depends on the sustainability of the processes used to produce these forms of energy, the broader EU commitment to renewable energies such as wind and solar are also relevant. The EU's Climate Change Package, adopted in 2009, calls for renewable energy to increase from about 8.5 per cent of the market to 20 per cent by 2020 (BBC 2009). Much of the implementation of these programs is delegated to member nations. The result is a complex and changing set of programs. As in the US, concerns over the sustainability of the production of biofuels, and their impact on food prices, has led to restrictions on how these count towards the targets, for instance through a substantial and increasing percentage of GHG saving relative to traditional fuels, double counting of more sustainable "second generation" biofuels that do not compete with food crops, and prohibition of the use for biofuels of land that contains "high carbon stock" or that has high biodiversity (Piebalgs 2009).[2]

Other jurisdictions: Public programs with private voluntary aspects

In Japan a "Top Runner" system was introduced in 1999, replacing less binding measures. This system set standards based on the best performing vehicles. Manufacturers were required to disclose information to consumers on fuel efficiency and CO_2 emissions, including a four-level ranking system established in 2004, and tax incentives to promote the purchase of vehicles with low emissions and high fuel efficiency were introduced in 2001 (Onoda 2008: 17). The Japanese system has been the most effective at reducing emissions. In sharp contrast to other jurisdictions, the average weight of vehicles has not increased. Also unusual is the tendency for firms to exceed the standards and to regard the government as lagging behind the industry in their commitment to fuel efficiency and emissions (Mikler 2009). For instance, by 2009 gasoline-powered passenger car fuel efficiency, at 16.9 km/liter, had exceeded the 2010 target of 14.8 km/liter.

These initiatives of firms can be considered a private voluntary aspect of Japanese vehicle emissions policy. Mikler has convincingly demonstrated the superior environmental performance of the Top Runner system of "co-regulation", with the private voluntary initiatives it encourages, as compared to the type of command-and-control evident in the CAFE system. In contrast to the CAFE system, where the industry successfully lobbied against an increase in the rigid standard, in the less adversarial Japanese system the industry is very actively involved in devising solutions. As Mikler shows, these differences are closely related to differences in the varieties of capitalism. Mikler extends his positive assessment of industry involvement in co-regulation to the EU case, and while the failure of the EU voluntary program, as discussed further below, should lead us to question this, it is certainly still likely that the scope for active industry engagement is indeed mid-way between the US and Japanese systems and has contributed to the correlation between that scope and the performance of the three systems (Mikler 2005, 2007, 2010).

In addition to emissions standards, the Japanese government has launched a series of environmental initiatives that support the transition to greener vehicles. These include the "Cool Earth 50" program in 2008, an Action Plan for a Low-Carbon Society, a Next Generation Fuel/Vehicle Initiative in 2007 and the 2006 New National Energy Strategy (JAMA 2009b: 11). A system of environmental performance certification labels on cars was introduced in 2004, and tax breaks for green cars have contributed to changing consumer behavior (JAMA 2009a: 5; JAMA 2009b: 10). The stimulus package that responded to the crisis of 2008 included subsidies to encourage consumers to purchase greener cars (Reuters 2009).

Other countries also display a diversity of approaches to controlling emissions. In Canada, amid growing concern about climate change and provincial initiatives that threatened to balkanize emissions regulation, the federal government indicated its intention to move towards stronger mandatory standards in place of the voluntary standards. Given the importance of the Canadian economy's links to the US, it is clear that Canadian emissions standards will be harmonized with those of the US. In Australia the voluntary targets call for an average of 222 g CO_2/km for cars and light trucks by 2010, almost double the EU target. Australian environmental organizations have called strongly for stronger mandatory targets, pointing to initiatives in other countries, but the industry has argued that Australian plans for integrating vehicle emissions into a larger emissions trading scheme are more advanced than elsewhere, and reliance on that scheme will only cost $25 to $30 per tonne of carbon saved rather than a $120 cost for technical improvements in cars (Porter 2009). China has developed new regulations that are intended eventually to match other standards, although considerable work needs to be done to implement them (An 2006; An and Sauer 2004: 15). China's official policies have been complemented by private voluntary efforts as well. The Auto Project on Energy and Climate Change (APECC) was established in 2004, and in 2006 was renamed and transformed into a non-governmental organization, the Innovation Center for Energy and Transportation (iCET). A variety of public and private actors are represented or involved in this think tank. Its programs include the development of the government-supported Energy and Climate Registry (ECR), an on-line voluntary mechanism for government agencies and businesses to report on their emissions.

Many countries are adopting programs to promote biofuels but production is concentrated in a few countries. In 2002 North America and Brazil alone accounted for 97 per cent of world ethanol production. Europe, China and Australia accounted for almost all of the rest. Europe accounted for 95 per cent of biodiesel, with North America accounting for the rest (IEA 2004: 30). The most successful ethanol program has been Brazil's. Ethanol accounts for 20 per cent of Brazilian car fuel (Hira and Guilherme de Oliviera 2009). A variety of government measures have been used, including setting targets for the percentage of ethanol in fuel blends, setting minimum prices for ethanol producers, providing credit, requiring the supply of ethanol at gas stations, and stabilizing the market through the use of strategic reserves.[3]

Biofuels involve difficult coordination challenges, not only between specific actors involved in the different stages of their production and use, but also in the regimes that must be mobilized for them to grow. These include, for instance, agriculture regimes, energy regimes and trade regimes (IEA 2004). Overall it is clear that, like other aspects of automobility, the way in which diverse actors, including business actors, are mobilized and coordinated will be crucial to the fate of biofuels.

Global public programs with private voluntary aspects

The global aspect of the vehicle emissions problem has contributed to the creation at the transnational level of programs showing the different shades of privateness in private voluntary programs. The International Energy Agency has devoted a great deal of effort to analyzing the technical and economic factors, and the associated economic costs and benefits, associated with alternative scenarios and, although it is a public organization, it has been active in establishing roadmaps for the transition from current technologies, an activity that is useful for coordinating across both public and private actors.

Although they are weaker than programs at lower levels, public authorities have also initiated programs at the global level with more significant private voluntary aspects. The International Council on Clean Transportation (ICCT) includes mostly public officials from the world's 10 largest vehicle markets, but also participants from civil society organizations and universities. An important genuinely public–private example is the 50by50 Global Fuel Economy Initiative (GFEI), which is a collaboration between the Fédération Internationale de l'Automobile Foundation (FIA Foundation), the International Energy Agency (IEA), the International Transport Forum (ITF) of the OECD and the United Nations Environment Program (UNEP) (GFEI 2010). The initiative aims to make cars 50 per cent more fuel efficient by 2050, worldwide. The FIA Foundation "was established in 2001 with a donation of $300 million made by the Fédération Internationale de l'Automobile (FIA), the nonprofit federation of motoring organizations and the governing body of world motor sport" (FIA Foundation 2010). The FIA Foundation is noteworthy as one of the few transnational civil society or consumer organizations that are playing an important role in the area of automobility and climate change. The ITF, while sponsored by the OECD, also includes high-level representatives from business and industry, research and civil society. The GFEI publications set out arguments for the importance of fuel economy in addressing climate change, and they provide various policy options and recommendations. The ambition of its vision is to combine the dissemination of information with more substantial advocacy: "the Initiative will engage with stakeholders at the global level, to get them to support and adopt the GFEI targets and at the regional and national level, to work on practical projects and programmes to implement the GFEI targets" (GFEI 2009: 17). It has over 90 partners from the public sector, industry, civil society and others (e.g. education, media).

A similar program with a somewhat different focus is the Partnership for Clean Fuels and Vehicles (PCFV), sponsored by UNEP. This partnership was initiated at the World Summit for Social Development in 2002, and especially aims to assist developing countries in improving the environmental performance of fuels and vehicles. One of its key instruments is a "Clean Fleet Management Toolkit", which is used in training sessions, and has resulted in strengthening of industry and government commitments to reduce automobile emissions (UNEP 2010).

The shift to a new technology like hydrogen will require very complex coordination of standard setting across a wide variety of industry segments. The primary transnational public body responsible for this is the public sector World Forum for Harmonization of Vehicle Regulations (World Forum), sponsored by the UN ECE in Geneva, also known as WP.29. The World Forum is also the leading body concerned with harmonizing vehicle safety regulations. It has worked on a roadmap for the development of hydrogen fuel cells which seeks to identify all the types of work on standards that is needed for developing hydrogen cars, including on-board storage safety, crashworthiness, electromagnetic susceptibility, recycling and fuel quality (World Forum 2004). It is clear that these standards will only be possible with input from public–private and non-governmental standard setters, such as ISO, Underwriters Laboratory (UL), the International Code Council (ICC), CSA (Canadian Standards Association) International, the US Fuel Cell Council (UNFCC) Codes and Standards Working Group, the US National Hydrogen Association (NHA), or the Society of Automotive Engineers (SAE), as well as ongoing university-based research from scientists and engineers. These standard setters are also working on alternative fuel systems other than hydrogen. Other relevant standards relevant to hydrogen that will need to be harmonized include fire codes, vehicle fuel systems codes, compressed gas codes, electric codes, building codes and mechanical codes.[4] There is a need to coordinate across alternative fuels as well. For instance the European Natural Gas Vehicle Association (ENGVA)[5] and the International Association for Natural Gas Vehicles (IANGV) have advocated for a worldwide harmonization of natural gas and hydrogen vehicle standards at the World Forum (ENGVA 2004).

This section has focused on the private voluntary aspects in programs initiated by public authorities. The importance of these private voluntary aspects has been evident: in a wide variety of settings and timeframes, from individual firms or local governments through to the global, from immediate production decisions to long-range planning, it is clear that the challenge of reconciling automobility and climate change cannot be addressed without the active voluntary engagement of private actors. At the same time, at all levels including the global level, public authorities continue to play a key role. When we turn to programs with greater degrees of voluntariness and privateness in the next section, this key role will become even more evident. It is notable that civil society and consumer organizations do not play a particularly strong role in this issue area, and this will be evident in the next section as well. This is due in part to the complexity of the technical questions and production processes that are involved, and also to the ongoing attachment of citizens to existing systems of automobility in which the car is central, or resistance to making the types of sacrifices that are needed to move to a new system.

Voluntary programs more fully managed by private actors

While it is impossible to discuss all the more purely private and voluntary programs that are relevant to automobility and climate change, it is useful to examine five of the more important and representative ones before turning to a more general analysis of the role of voluntary programs in this issue area as a whole. The program with by far the greatest degree of privateness, voluntariness and programness is the one agreed between the EU and the European auto industry in 1998. As discussed below, even in this case public authorities played a critical role in the designing, joining, monitoring and compliance aspects of this program, and we may call it a negotiated arrangement to distinguish it from the greater degree of government control in the arrangements discussed in the previous section.

The EU program negotiated with the auto industry was launched amid a high level of enthusiasm in the EU for voluntary programs in environmental and other issue areas, particularly evident in the 5th Environmental Action Program. Instead of developing legislation or regulations, the EU relied upon a commitment from the European Automobile Manufacturers' Association (ACEA) to meet a target for tailpipe carbon emissions. As the largest and most important private voluntary program for implementing carbon emission reductions, it deserves particular attention. In this section a second voluntary program launched by Canada will also be examined. A third important private voluntary program that is discussed is the World Business Council for Sustainable Development's (WBCSD) Sustainable Mobility Project (SMP), which issued two reports setting out the industry perspective on longer-range sustainability planning issues. The section concludes with a fourth type of voluntary program run by the International Organization of Motor Vehicle Manufacturers (OICA), the global association of vehicle manufacturers.

The ACEA, KAMA and JAMA negotiated voluntary programs

In 1998 the European Commission agreed with the ACEA that the ACEA and its members would commit to a target of 140 g CO_2/km by 2008, a 25 per cent improvement, with an interim 2003 target of 165–175 g CO_2/km. The agreement stipulated that the Commission would negotiate similar agreements with the Korea Automobile Manufacturers Association (KAMA) and the Japan Automobile Manufacturers Association (JAMA), and this was done, but with an extra year for those manufacturers to reach the 140 g CO_2/km target.[6]

The voluntary agreements were the first part of a three-pillar EU strategy. The second pillar was an EU Directive requiring labels on new cars that would specify fuel efficiency and carbon emissions. The third pillar was a commitment by the Commission to make stronger use of fiscal measures to encourage consumers to purchase cars with lower emissions.

The agreement called for yearly monitoring of the progress made towards reaching the targets. This monitoring was to be carried out jointly by the ACEA and the Commission. The Commission declared its intention to pursue mandatory rules, should sufficient progress not be made under the voluntary agreement. In 2007 the Commission announced that the reductions were not occurring fast enough and that as a consequence it would move to replace the voluntary program with binding rules. Of particular concern was the EU target of 120 g CO_2/km by 2012, which the EU saw as essential to meeting its Kyoto commitments.

A number of factors contributed to the abandonment of the voluntary agreement. By 2004 average emissions had only been reduced to 163 g CO_2/km, a 12 per cent reduction, and it was clear that the targets were not going to be met. Most of the reductions had come from technological advances rather than the effects of labeling or fiscal measures. A major problem was that technological advances had been offset by a growth in the size and power of the cars that were being sold in Europe. Some of this growth may be attributable to the greater profitability to the industry of the more powerful cars, and the industry's use of images of power and speed to market cars. However, it also reflected the failure of public authorities to implement the type of labeling or fiscal measures that might have encouraged consumers to favor greener cars. When analyzing these voluntary agreements, interpretations of their success or failure are dependent on the goals to which they were harnessed, as well as assumptions about counterfactuals. Supporters of the agreements can point to the reductions that were achieved. Critics argue that the reductions were not significantly different than business as usual, and that the severity of the climate change challenge requires far deeper cuts in emissions than these voluntary agreements were able to achieve.

A closer look at the designing of these private voluntary programs helps explain their fates. This history makes very clear that these private voluntary programs would not have been developed without the initiative and power of public authorities, but this public involvement came with its own problems. The first such problem was the clash between governments when they seek to promote the interests of their own country's firms. The second such problem was the failure of public authorities to act more aggressively in developing their own regulations and in the negotiation of the private voluntary agreements.

The design of the voluntary agreements can be traced back to the formation in the mid-1980s of the Motor Vehicle Emissions Group (MVEG), composed of national officials and auto industry representatives (Keay-

Bright 2000: 17). One problem with the proposals advanced in this group was their failure to overcome national interests. This was evident, for instance, in the contrast between the German preference for differentiating limits by car size, which would mitigate the burden on the German industry's more powerful cars, and the French preference for an absolute limit, with corresponding penalties for those exceeding it and rewards for those coming in under the limits. The French industry produces small cars that would benefit from this arrangement. A second type of problematic proposal was any that called for the use of fiscal measures. Taxes high enough to bring about the necessary reductions would be highly unpopular among voters. Moreover, it was not clear that the European Union had the political or legal capacity to bring about this type of coordinated fiscal action. A third type of problematic proposal was the idea of tradable permits. These were ultimately seen as too complex to be practical. Clashes within the Commission, especially between DG3 (internal market and industrial affairs) and DG11 (environment), further complicated discussions, with DG3 being particularly supportive of voluntary approaches (Keay-Bright 2000: 19). The idea of using private voluntary programs, complemented by relatively modest labeling and fiscal measures, appeared to offer a way around these problems.

The negotiation of the private voluntary program with ACEA clearly displayed the ongoing importance of clashing interests. The difficulty for the intensely competitive auto companies of collaborating, even given their relatively small numbers and oligopolistic control of the industry, illustrates the way in which the small size and similarity of groups of powerful but competing actors can make it harder rather than easier to support voluntary collective action. ACEA's initial reaction to the idea of a voluntary agreement was not particularly supportive, as it would have preferred to avoid direct restraints on its members' manufacturing processes (as compared, for instance, to no restraints, or revenue-neutral fiscal measures aimed at altering consumer preferences). In 1997 ACEA offered to commit to a 10 per cent reduction in exchange for commitments from political authorities on research funding, safety standards, and other matters (p. 23). Keay-Bright notes that the inability of ACEA members to agree on how to apportion their own commitments contributed to the modesty of their offer (p. 24). Ultimately the stronger commitment to a 25 per cent reduction resulted from the strengthening of the resolve of the public authorities.[7] Negotiations shifted up from under-resourced officials within the Commission to the Commissioners for the Environment and for Industry. At the same time, the failure of public authorities to begin to prepare legislation that could be brought, in the event that the voluntary agreement did not work, likely undermined the effectiveness of the agreement. Concern about competition rules also restricted the degree to which the Commission or the industry were able to establish firm-specific targets, and instead the vaguer notion of "equivalent effort" was used (Keay-Bright 2000: 44).

The other three elements of private voluntary programs highlighted in Chapter 1, joining, monitoring and compliance, were also, clearly, heavily influenced by the actions of political authorities. ACEA insisted that its competitors be brought under similar agreements, but those agreements with KAMA and JAMA were negotiated by the Commission. ACEA also wanted the EU to try to extend the commitments to car sales in other countries, but doing so was far beyond the capacity of industry associations and would have involved interstate negotiations that were not politically feasible. In the end the agreements negotiated by the Commission covered all significant automobile sales in the EU: in 2002 ACEA, JAMA and KAMA accounted for 99 per cent of car sales in the EU, with the remainder coming from smaller, specialty manufacturers (Institute for European Environmental Policy 2003: 5).

Monitoring and compliance were similarly dependent on the relation between private voluntary agreements and the public authorities. Whether or how ACEA was monitoring and enforcing compliance among its members was not clear. Even well into the process, interviews by Keay-Bright (2000: 44) indicated that ACEA members "have still not discussed the contributions expected from each other ... According to ACEA it is not possible to define 'equivalent efforts' so contributions will be defined on an ad hoc basis." A 2005 World Resources Institute/SAM Group report was strongly critical of the refusal of the industry to make public the commitments or the progress of individual companies, arguing that this made it impossible for investors to incorporate financial risks and other relevant information into their decision-making (Sauer, Mettler, Wellington and Hartmann 2005). This reveals the absence of the type of firm-specific private benefit that can be important in private voluntary arrangements, as noted in Chapter 3 by Hsueh and Prakash. Ultimately both monitoring and compliance were reliant on the yearly joint action of the industry and the Commission, with a yearly report to the European Parliament and Council, as specified in a 2000 Decision (Sauer et al. 2005: 48). The weakness of the monitoring and compliance mechanisms certainly contributed to the failure of the agreement to meet its targets.

The Canadian voluntary programs

While the European voluntary programs are especially important, Canadian voluntary programs displayed similar problems that merit some consideration here. In Canada, which is heavily integrated into a North

American production system for automobiles, fuel economy standards have been driven by the US CAFE regime. However, in 1996 the federal and provincial Canadian governments decided to pursue a voluntary strategy towards climate change through a "Voluntary Challenge & Registry" (VCR), which had been initiated by a multisector working group in 1995, and this was significant for the automobile industry both through direct involvement and through participation of other industries that are involved in broader well-to-wheels aspects of GHGs. "In October 1997 VCR completed its transition from a government incubated program to a stand-alone private-public partnership" (VCR 2005a, 2005b). Two-thirds of VCR funding was private and one-third public. It adopted a structure involving a "Council of Champions", with representatives from various sectors and regions, together comprising over 78 per cent of business and government operations, to reduce GHGs in Canada, and including the auto sector. Industry members of a board of directors were elected from an annual meeting of the Council of Champions, and the government members were appointed by the Minister of Natural Resources Canada. A condition of membership was to report on GHG emissions at a "Champion level". The VCR developed relationships with a wide variety of associations, many of which are relevant to automobility, either directly or through their involvement in the production of energy resources, such as nuclear power, that could be converted into energy carriers such as hydrogen or batteries. Examples include the Canadian Association of Petroleum Producers (CAPP), the Canadian Vehicle Manufacturers Association (CVMA) and the Greenhouse Gas Emissions Reduction Trading Pilot (GERT). Taken as a whole, these relationships provide an interesting illustration of the way that established associations and newer arrangements tend to both play roles in addressing new complex issues such as climate change.

By 2002 the VCR was boasting of its ability to publish information on GHG emissions: "Hundreds of organizations have registered reports annually, each describing GHG emissions, reduction projections, target setting, measures to achieve targets, and/or results, as well as education training and awareness activities related to GHG emissions reductions" (VCR 2002). It proudly noted the large increase in visitors to its website. The coordinator of the New Directions Group, an informal forum involving business and civil society representatives, applauded the voluntary approach in 2002: "What we're talking about is a voluntary agreement between business and government for the management of greenhouse gas emissions for which there are very real performance expectations and very real consequences if you don't make it, and very real rewards if you do" (VCR 2002). In theory this aimed to provide the type of strong monitoring and compliance mechanisms that Hsueh and Prakash, elsewhere in this volume, identify as important for the success of private voluntary programs.

However, by 2002 it had also become apparent that the restraints on emissions associated with the voluntary program were modest at best. A 2002 report by the Pembina Institute and the David Suzuki Foundation, which reviewed the data submitted to VCR, noted that "our findings show that voluntary action taken by most of Canada's industrial firms to cut greenhouse-gas emissions have been wholly inadequate" (Bramley 2002: 21). Only 102 of 493 industrial entities registered with VCR had actually reported their 2000 emissions (Demerse and Bramley 2008). While the head of VCR rejected the findings, his defense claimed that companies would have had far greater emissions without the VCR, and that because the businesses are growing rapidly it is hard for them to find energy-efficient measures large enough to offset those increases (Middlestaedy 2002). In short, there seems to be little dispute that the voluntary measures were not bringing about absolute reductions in emissions. The dispute revolved around the wisdom of seeking those absolute reductions and the timeframe within which absolute reductions or reductions in emissions intensity would be achieved (Jacard and Bataille 2002).

By 2005 the Government of Canada had abandoned its voluntary approach to GHG reductions in recognition by both government and key industry actors that compulsory regulation would be needed to achieve more aggressive reductions. A draft Canadian government document leaked in 2005 noted that the "voluntary approach and limited incentives [are] not sufficient to drive substantial change". Policymakers would "need more consideration of regulation and taxation to drive behavioral change and technology development and uptake" (Lucas and Potes 2006: 1).[8]

In Canada fuel economy standards, since 1976, have been governed by a voluntary agreement by manufacturers to meet the US standards. The Canadian Company Average Fuel Consumption (CAFC) targets translate CAFE standards into the Canadian equivalents. Despite the move away from voluntary programs evident in its abandonment of the VCR, in 2005 the Government of Canada signed a new Memorandum of Understanding with the Canadian automotive industry that called for the latter to take voluntary actions to reduce GHG emissions for light duty vehicles by 5.3 megatons by 2010, representing about a 25 per cent reduction. Careful independent analysis of the effects of the agreement reveals that it most likely did not improve performance beyond business as usual (Lutsey and Sperling 2007). In 2007 the government announced that it would abandon the voluntary agreement when it expired, and the long tradition of voluntary agreements that preceded it, and replace it with mandatory legislated standards: "This government has decided to use regulation over voluntary agreements as it provides a greater degree of certainty, predictability and accountability" (Government of Canada 2007: 1).

The long history of voluntary fuel economy programs in Canada has been shaped by the integration of Canada's auto industry into the US industry. Since manufacturers mostly produce to a single standard to export throughout both markets, the US standards in effect were mandatory for Canadian manufacturers, and the voluntary agreement had little independent effect. Once the government began to speak of other more aggressive standards, in 2005, it quickly moved away from the voluntary approach.

By 2008 the commitment of the Canadian government, under the Conservative Party, to achieving absolute reductions in GHGs such as those included in its Kyoto obligations had also significantly weakened. The shift from voluntary to mandatory standards, therefore, will likely not lead to a dramatic strengthening of Canadian regulation of GHG emissions in the short or medium term.

Overall the Canadian voluntary programs were even less successful at addressing the problem of automobility and climate change than were the European schemes. The Canadian program displayed similar characteristics with regard to design, in the degree to which it was initiated by public authorities. The VCR membership was multi-industry, and this diminished its ability to focus on automobile emissions more specifically. Monitoring and compliance was based on reporting, and even the weak threat of legislation that the European Commission could wield was absent. The fuel consumption agreements were entirely dependent on US CAFE standards.

World Business Council for Sustainable Development's Sustainable Mobility Project

The SMP, referred to earlier, was an ambitious private voluntary program designed to contribute to long-range planning for the sustainability of transportation systems, focusing especially on road transportation. Its members held senior positions, mainly at the CEO or board chair levels, in GM, Toyota, Royal Dutch/Shell (these three co-chaired the project), BP, DaimlerChrysler, Ford, Honda, Michelin, Nissan, Norsk Hydro, Renault and Volkswagen. As the members noted, "Normally our companies compete vigorously, so to produce such an in-depth, agreed analysis is a distinct accomplishment" (SMP 2004: Foreword). The SMP was the largest member-led sector project ever undertaken by the WBCSD (Stigson 2004). As the WBCSD President noted, "The project fostered unprecedented cooperation among a core group of major companies representing vehicle technologies, fuels and parts suppliers. In total, the group represented over three quarters of the production capacity of motor vehicles globally" (Stigson 2004). The SMP's first report, which assessed the existing state of mobility worldwide, was issued in 2001, and its second, more future-oriented report was issued in 2004.

To discern the SMP's distinctive contribution as a private voluntary program it is useful to compare it to the 2009 International Energy Agency report *Transport, Energy and* CO_2*: Moving Toward Sustainability*, since both reports have similar purposes and cover similar topics, but the IEA is an intergovernmental organization. There are two main differences that stand out. First, the SMP reports have a much broader conception of sustainability. Second, the SMP recommendations and conception of what is possible are more modest. It is useful to briefly look at each in turn.

With regard to the two conceptions of sustainability, the 2004 SMP report sets out 12 indicators of sustainability. One of these is GHG emissions. Another is impact on the environment and on public well-being, which includes conventional emissions other than GHGs, the impact on ecosystems, and noise. A third is resource use, which includes energy use and security, land use and materials use. Other indicators are some that might not typically be included in some conceptions of sustainability, including accessibility, financial outlay required of users, travel time, reliability, safety, security, resource use, equity implications, impact on public revenues and expenditures, and prospective rate of return to private business. By contrast the IEA report focuses very heavily on CO_2 emissions. This undoubtedly reflects the specific competence of the IEA in energy matters, and a decision by public authorities to focus on CO_2 in this particular publication, although it should be noted that the IEA's parent organization, the OECD, specializes in "whole of government" approaches to problems, and the IEA consulted well beyond energy experts in its production of its report.

However, the two different treatments of sustainability also reflect the strengths and weaknesses of private voluntary programs relative to public sector initiatives. One strength of the SMP is its recognition of the broader connections and purposes of road transport and mobility more generally. An overly narrow focus on CO_2 reductions can obscure the trade-offs that might be involved in pursuing those reductions. These trade-offs include a loss of mobility, including in ways that exacerbate the equity-related "mobility opportunity divides" that SMP research highlighted. These divides include those between developing and developed countries, but also mobility problems experienced by socially excluded groups. They also include the trade-off associated with safety measures, which can add weight to vehicles and thereby increase emissions. It is also certainly useful to include other ways of reducing vehicle emissions that do not rest so heavily on vehicle technologies, such as measures to mitigate congestion, or car-sharing programs. Although the capacity and reach of governments typically might lead us to expect them to integrate a wider variety of policy-relevant factors in their programs than would a private voluntary program, it is also the case that governments can make a more single-minded pursuit of a narrow goal because they need to simplify to reach political agreements, and because their bureaucratic structure inhibits their responsiveness to complex environments, including the specific properties of production processes or the preferences of the population at large. The breadth of the SMP approach to sustainability reflects these advantages that stem from its character as a private voluntary program.

At the same time the breadth of the SMP approach expresses the business interests of its members in ways that do not match up with the public interest in addressing climate change. Some of the non-conventional indicators of sustainability that it has included are some that industry has performed particularly well on, like the improvement of safety or the reduction of noise, or those that fit well with the industry's commercial interests, like public investments to reduce traffic congestion and to improve roads, improving the competitiveness of road transport relative to other modes, or with the industry's sales interest in expanding access to mobility in the developing world. The SMP reports also emphasize the positive aspects of industry, such as the contribution of transportation to economic growth. Overall an effect of the SMP approach is to diminish the greater intensity of public authorities' focus on CO_2 reductions. The single-mindedness of governments can be an important asset in addressing challenges that require much faster, stronger and focused action than industry is comfortable with or able to initiate.

The second difference between the SMP and IEA approaches is the more modest ambitions of the former in setting out plans for reducing CO_2 emissions. The IEA report sees a multimodal effort as able to achieve a 40 per cent reduction in CO_2 emissions below 2005 levels by 2050, with transport oil use being cut in half. The 2004 SMP report, in contrast, sees even its most ambitious scenario, involving the adoption of new technologies and fuels, as returning 2050 CO_2 emissions from road vehicles to their 2000 level (SMP 2004: 117). Some of the differences in these scenarios are due to the inclusion in the IEA projections of a shift in modes from CO_2-intensive modes to less intensive ones (such as a shift from cars to bicycles, or a shift in freight transport from truck to rail). However, even if this modal shift is excluded, the IEA estimates that a 30 per cent reduction in transport CO_2 emissions from 2005 levels by 2050 can be achieved for a cost of less than US$200 per tonne of CO_2 saved (IEA 2009: 31). The 2004 SMP report's most ambitious projection envisions GHG road transport emissions in 2050 as about 1 Gt per year above their 2000 level (SMP 2004: 118). By contrast the IEA report anticipates reductions by 2050 in light duty vehicles of 5 Gt per year below their 2005 levels (IEA 2009: 36). It should be acknowledged that differences between the methods and variables used in the two studies make a precise comparison impossible. Nevertheless it is clear that the SMP projections are very significantly less ambitious than the IEA projections.

The 2004 SMP report acknowledges the need for initiative on the part of public authorities if GHG emissions are to be made sustainable: "The benefits of GHG emissions reduction accrue to society at large rather than to any individual transport user. So the incentive for individuals to incur significant extra costs voluntarily to acquire and operate vehicles that emit significantly fewer GHGs is likely to be quite limited. Incentives will probably be needed, and only governments have the resources and authority to create them" (SMP 2004: 105). While further improvements in existing technologies are foreseen as relatively unproblematic, the report is cautious about a shift to what it sees as very uncertain new technologies and fuels: "All these uncertainties mean that it is not practical at this stage to define what incentives might eventually be needed to enable their widespread use. However, it is possible to describe what governments can do over the next several years to help industry reduce these uncertainties to the point where a meaningful discussion regarding implementation can be held" (SMP 2004: 105). The differences between the two reports illustrate the degree to which the shorter time horizon of private actors makes it difficult to address longer-range problems such as climate change, and to reduce uncertainty about how to proceed. Business experiences uncertainty as a deterrent to the large-scale, long-range investments that are needed to address climate change, and experiences limits in its ability to manage this uncertainty, and public authorities must then be relied upon to reduce this uncertainty.

In terms of the four elements of private voluntary programs, the SMP illustrates the complexity of the role played by such programs in this sector. The program was not designed to bring about compliance, and consequently monitoring was seen as unnecessary.[9] The design and membership of the SMP was initiated by its members, facilitated by the WBCSD, and seems to have struck a reasonable balance between inclusion and efficiency, since key players, accounting for three-quarters of automobile production capacity, were represented at a very high level, and a much broader inclusion of other actors would have inhibited agreement.

The Organisation Internationale des Constructeurs d'Automobiles contribution

The Paris-based OICA is the main transnational association that represents auto manufacturers. Traditionally it has been relatively weak and has done little other than lobbying for the costs of climate change to be shifted to actors other than the assemblers, for instance through the efforts of its support for the World-Wide Fuel Charter to shift these to the oil industry and coordinate fuel standards with car production technologies (Porter 2002). It has also provided some commentary on regulatory developments at the intergovernmental World Forum for Vehicle Harmonization. Its weakness is a reflection of the oligopolistic character of the industry and its organization on a regional basis. Firms find it efficient to lobby national governments directly or to work through national or regional associations. The membership of the OICA includes 41 national trade associations from around the world.

A major emphasis of the OICA is to stress the degree to which the auto industry is only one source of CO_2: "It is therefore essential to find the measures that will deliver the biggest CO_2 reductions for the lowest cost to society, and concentrate first on those . . . The most cost efficient approach must be used across all sectors if society is to effectively meet the challenge of climate change" (OICA 2010: 7). Like the SMP, OICA's integrated approach includes initiatives in which the cost/benefit ratio for the industry would be favorable, such as government funding of R&D and infrastructure, government incentives for fleet renewal, congestion mitigation and consumer eco-driving.

The Worldwide Fuel Charter (WWFC) was first established in 1998 to "promote greater understanding of the fuel quality needs of motor vehicle technologies and to harmonize fuel quality worldwide in accordance with engine and vehicle needs" (OICA 2006: i). The members of the WWFC include vehicle manufacturers' associations from around the world. The fourth edition of the WWFC, issued in September 2006, is a highly technical document, and even its list of comments and responses from the consultation conducted before the document was issued are mainly about technical matters. In 2009 the WWFC Committee issued similar guidelines for biodiesel and ethanol (WWFC Committee 2009).

Overall the OICA and the WWFC are mainly concerned with disseminating information about the auto industry and facilitating lobbying that is primarily occurring elsewhere. Given the need to involve the industry in the types of activities that the OICA is involved in at the World Forum, its contribution is not entirely inconsequential, even if it is only one of many business actors working on similar issues. The Worldwide Fuel Charter, however, is more uniquely consequential. The power of the vehicle manufacturers and their ability to collaborate on these standards gives the standards a lot of force and will help coordinate the move to greener fuels. There will inevitably be conflicts between the vehicle manufacturers and the fuel producers over which should bear the costs of adjusting to change, and the WWFC plays a part in these. However, it is unlikely that either the OICA or the WWFC will independently motivate change significantly beyond a business-as-usual scenario.

Other private voluntary programs

There are a vast number of other private voluntary programs that can be expected to play some part in addressing the relationship between automobility and climate change. The complexity of the number of technologies and actors involved in moving towards greener forms of mobility, the uncertainty associated with the transition to new technologies, obstacles to the flow of information, and the need to coordinate at all scales, create a great number of types of collective action that are needed. The functions of these voluntary programs can include lobbying, standard setting, production or sharing of technical standards, the coordination of roadmaps for technological trajectories, among others. The associations and programs display a type of overlapping pluralism that is inevitable in an issue area of this complexity, but that also makes collective action difficult.[10]

At times these private voluntary programs work against the greening of automobility. For instance, in the US the Coalition for Vehicle Choice claimed to be a non-profit representing 40,000 member groups and individuals, but was created by the industry to fight against fuel efficiency standards. Similarly the Global Climate Coalition (GCC), representing industries reliant on fossil fuels, including the OICA, aggressively lobbied against the idea that emissions caused global warming (Revkin 2009). At other times, by promoting marginal improvements in existing technologies, the industry can help shift programs away from the more aggressive transformations that may be needed to address climate change.

Some of the effects of private voluntary programs are not intended to address climate change directly at all, but nevertheless can be relevant. For instance, programs that promote the use of sound systems and other on-board consumer electronic devices, such as the automotive multimedia interface collaboration (AMI-C), can reduce sustainability by adding weight to vehicles, or aid sustainability by fostering the use of intelligent transport systems. Climate-relevant biofuel and hydrogen programs can include many applications other than vehicles.

Despite the negative impacts that some private voluntary programs can have on sustainability it is clear that the coordination that is needed to move forward would not be possible without such programs.

Conclusion

The complexity and challenges that the sections above have revealed make it clear that entirely autonomous private voluntary programs are not likely to play a significant role in the global effort to address the contribution of automobility to climate change. The four institutional conditions that were identified in Chapter 1 help explain this.

The first set of institutional conditions pertains to the interactions and initiatives of individual firms. In the case of automobility, the oligopolistic structure of assemblers and their dominance over suppliers strongly shape its capacity to address climate change. The oligopoly stems in part from the character of the industry's production process, which requires certain economies of scale and internalization of tightly coordinated production processes. This gives assemblers such as GM or Toyota considerable potential for developing climate-friendly technologies such as hybrid engines and imposing these on their suppliers. However, such potential is offset by the intense competition between assemblers. This has contributed to the relatively unimpressive response of individual firms to the climate change challenge. The complexity and uncertainty of the transition to a more climate-friendly form of automobility requires extensive coordination, and although some of this is provided by industry associations, the oligopolistic structure of the auto industry, where powerful firms take care of their own problems individually, has contributed to a relatively weak associational capacity.

The second conditioning factor was the role of public institutions, and this chapter confirmed the emphasis in Chapter 1 on the huge impact these can have. In all jurisdictions discussed above there would have been very little forward progress without public initiatives. As anticipated by Chapter 1, the EU delegation of emissions reductions to the industry was motivated in part by weaknesses and failures in the EU institutions. Part of the failure of the voluntary programs can be traced to the degree to which they relied too heavily on well-intentioned but under-resourced expert officials in the Commission, who found themselves challenged in negotiating with the industry. In contrast, the greater capacity of the Japanese government to work productively with industry contributed to the success of the Top Runner model there. This model also confirmed Chapter 1's emphasis on the ability of business to take initiatives of its own, as did other initiatives ranging from the transnational World Business Council for Sustainable Development's Sustainable Mobility Project to the more focused technical standards of the European Biodiesel Board.

The third conditioning factor discussed in Chapter 1 was the role of civil society. The introduction pointed out that civil society was more effective in promoting general awareness of the problem than in engaging with the type of specific and detailed programs of the type needed to move the automotive industry to a greener technology. This chapter confirmed this. Partial exceptions include the Fédération Internationale de l'Automobile Foundation's important support of the 50by50 Global Fuel Economy Initiative, and the role played by civil society in criticizing the poor performance of the voluntary programs in the EU and Canada. Some parts of civil society have also certainly challenged automobility locally, such as proponents of cycling or opponents of expressways. However, consumer preferences for larger, more powerful cars have been a major problem in both the US and the EU, indicating that the initiatives of citizens are not always positive with regard to the climate. Rather than assuming that civil society will pressure business to be greener, we may say that civil society has a diffuse and ambiguous impact, and it, just as much as business, still needs to be further mobilized in support of alternatives to our current system of automobility.

The fourth conditioning factor was various scientific expert communities. The role of such expertise is crucial in addressing the problem of automobility and climate change. Technological solutions that allow the existing industry to shift to greener transport while maintaining their profitability and providing consumers with vehicles that resemble existing cars in their power and performance are, unsurprisingly, the ones that are most favored by the industry and its supporters in government. Politically, these solutions are far easier to promote than those that require industry or consumers to make sacrifices. It is in the self-interest of industry to shift to new technologies since it recognizes that sooner or later petroleum-based engines will be unsustainable. However, even more ambitious programs than those preferred by the industry will also require high levels of innovation since fully abandoning the car is not feasible politically or practically. The mobilization of scientific expertise has most prominently involved public–private partnerships such as FreedomCar, in which the broader scientific capacity that governments are able to mobilize can be matched with the more specific, focused engineering and commercial expertise that the firms possess. In this industry the scientific communities, like civil society, have a diffuse and ambiguous impact. Much scientific expertise is mobilized to make incremental changes in the existing technology, which will be inadequate for addressing the larger climate problem, and overall automobile-related scientists do not play the type of role that the IPCC does in continually alerting everyone to the urgency of the problem.

Chapter 1 also identified designing, joining, monitoring and compliance as four functional requirements for successful private voluntary programs. The automobile-related programs discussed in this chapter illustrate the difficulty of identifying and carrying out these functions with a problem as complex as automobility. In an ideal typical model, a set of firms might get together and propose to coordinate their joint efforts to address climate change, designing their program, deciding on their membership, and putting mechanisms in place to monitor and enforce their own compliance. However, in this sector most private voluntary programs are quite different than this ideal type. Some public or private programs might provide some parts of some functions, while others may complement these with other functions. It is useful to comment briefly on how this is evident with each of the four functions.

The massive coordination challenge and high level of uncertainty involved in transitioning to new modes of transportation means that the designing function often cannot simply involve the design of a standalone organizational arrangement to help firms address a predetermined problem. Not all programs are designed to result in compliance with a fixed set of rules, nor should they be. Some valuable programs instead require a strong commitment of resources from business simply to identify sustainable strategies or technologies for the industry, as exemplified by the WBCSD's Sustainable Mobility Project. Overall the designing function operates at multiple scales, ranging from conceptualizing how automobiles, fuels and transportation infrastructures can be altered simultaneously and quickly over long time periods, to smaller-scale projects such as how technically to guarantee the quality of biofuels at the pump. In some cases the fragmentation of designing functions and their separation from the other three functions can be a weakness, but can also be appropriately decentralized responses to a challenge this complex.

Chapter 1 noted that the impetus for the design of a program can come from multiple places, sometimes modifying existing organizations and sometimes creating entirely new programs. This is certainly reflected in the automobile case. For instance, the Worldwide Fuel Charter arose from the OICA, which itself is a federation of national automobile manufacturers' associations. The SMP arose from the WBCSD and the initiatives of major private-sector players in the transport sector. The FreedomCar and Fuel Partnership program, even though it was initiated by government, has fostered much closer collaboration between industry partners. The biofuel associations are relatively new clubs that did not grow out of older automotive associations, although their design draws on existing associational practices.

The joining function displays similar complexity. Where the purpose of a program is to develop a roadmap rather than enforce compliance with existing rules, it is more important to get representatives from all the relevant industry sectors rather than all the firms in any particular sector. So, for instance, the SMP included most but not all of the major assemblers, but also Royal Dutch/Shell, BP, Norsk Hydro and Michelin. The multiple hydrogen codes mentioned above are most likely to be managed by organizations that are not specific to the automotive industry and do not need to have all firms in the industry join. In contrast, the EU voluntary emissions program would have failed even more quickly if it had not included all the assemblers, since those not included would have enjoyed a competitive advantage that would have been intolerable for members.

The monitoring function also varies widely. Programs that are initiated by public authorities with private participation have typical reporting and review procedures and are the target of critical scrutiny from activists and scholars, as evident with FreedomCar. The Canadian voluntary program involved public reporting expectations, but compliance with these was unimpressive. The EU voluntary program only required reporting of individual firm performance to the association, which reduced public accountability for poor performers. The EBB biodiesel testing involves detailed mandatory technical procedures for collecting and analyzing samples. When the purpose of the program is planning rather than regulation, monitoring may not be important. In many cases the absence of private monitoring mechanisms simply reflects the lack of development of specific climate change-related rules. For instance SAP, a firm that provides private compliance assessments for sustainability regulation in the automotive industry, emphasizes chemical regulation, product safety, health and safety management, and other risks, but makes no direct mention of climate change (SAP 2011). A major challenge in monitoring is a lack of consensus about the measures that are needed and how speedy the implementation of those measures must be. In another period the emissions reductions that were achieved by the EU voluntary program may have been judged a great success, but in relationship to the reductions that European governments now agree are necessary, they failed. Despite these caveats, there are significant examples above of clear targets where monitoring revealed that they would not be reached, such as the hydrogen goals of FreedomCar, and the emissions reductions of the EU and Canadian programs.

Where compliance with tough rules is required for a program's success, there is no example of this being achieved through purely private mechanisms. Many firms will voluntarily comply with technical standards to enhance the interoperability and credibility of their products, and many firms will voluntarily comply with the mapping or information sharing programs because these can be beneficial for the individual firm. However, the harder adjustments, involving net costs and competitive disadvantages, such as the voluntary emissions reductions schemes, failed to achieve compliance by private means. Part of this is due to the lack of private pressures that could provide an incentive for compliance. We are not yet at a catastrophic time for an industry, such as if oil reserves were exhausted. The threat to the industry is too far in the future for it to voluntarily incorporate difficult change into its operations. Consumers are sending ambiguous messages about their desire to mitigate climate change.

Looking at the record overall in this sector of the privateness, voluntariness and programness of the initiatives that have been taken to address automobility and climate change, the predictions that were drawn from literatures on global governance and regulation and that were set out at the beginning of this chapter are supported.

The first prediction was that that the state's unique capacity to manage binding commitments across multiple settings and long timeframes involving large numbers of actors would be necessary for the aspects of problems for which private and other smaller-scale institutions are too weak or too small. None of the private voluntary programs examined came near to the scale or capacity to manage the major technological changes that are needed. The most ambitious program, the European voluntary programs, failed, and public authorities needed to step back in. The SMP, while bold in its vision, acknowledged the need for governments to take the lead.

The second prediction was that private voluntary programs will be especially important for addressing smaller-scale, more focused, detailed problems in rapidly changing environments. The above discussion has made clear that wherever detailed application of policies to specific processes is needed, industry has been and must be involved, and for this to be successful this involvement needs a significant degree of privateness, voluntariness and programness. The complexity, uncertainty and interdependence of the technologies involved require active independent engagement of industry actors. Even in bold system-changing initiatives such as FreedomCar, the active independent engagement of private actors is important in ensuring that they do not veer off in directions that reduce the likelihood that overall goals will be met. The early excessively heavy emphasis of FreedomCar on hydrogen technologies might have been avoided if more careful attention to the constraints present in the industry itself had been considered.

The broader range of factors that the SMP reports considered, as compared to the IEA reports, provide an important counterweight to the tendency of governments to focus excessively narrowly on particular goals.

This narrowness of the focus of public authorities, which comes from the properties of international negotiations, the need to be decisive in the face of complexity, or the distance of public authorities from the complexity of the industry, was not something that was predicted by existing literatures, which instead stress the breadth of the capacity and perspective of public authorities relative to private actors.

A third prediction that was drawn from existing literatures was that, when shifts in policy paradigms are needed, neither the state nor business by themselves can bring about this change, and broader public involvement in norm change is needed. It is clear from the above discussion that the failure to adequately address the problem of automobility and climate change is not only due to industry resistance. Governments too can fail to act in an optimal manner. Civil society engagement is not a panacea, since many consumers favor high-emissions cars, but this confirms that to bring about the scale of change that is needed, the public must both alter its own practices and put pressure on government and business to do so as well.

Overall we may conclude that both voluntary private programs and mandatory public programs are needed, and that they must complement one another to be successful. Government must provide leadership on the broader framework and compliance mechanisms, and business must provide the specific initiatives and knowledge that are needed to make change practical. The complexity and severity of the climate-related challenges facing the automobile industry cannot be solved by either government or business acting too independently of one another. In order to achieve the complementarities that are needed it is crucial to continue to consider the types of conditions for success and failure that this chapter has discussed.

Notes

1. USCAR is an umbrella group for collaborative research involving Chrysler, Ford and General Motors. In 2003 the major energy companies (BP America, Chevron Corporation, ConocoPhillips, Exxon Mobil Corporation and Shell Hydrogen LLC) were brought into an expanded FreedomCar & Fuel Partnership.
2. Andris Piebalgs: Member of the European Commission responsible for Energy.
3. Elsewhere biofuels are not as well developed but new production capacity is being debated or put into place. For instance, in sub-Saharan Africa the role of private actors varies across countries, with Nigeria and Uganda relying more heavily on government, with South Africa, Tanzania, Zambia and Malawi relying more on the private sector and civil society actors. There are numerous challenges in achieving the scale and efficiency needed if African biofuels are to become competitive with gasoline, including the need to coordinate across many actors, including risk-averse farmers, those responsible for pressing the oil from plants, the suppliers of conventional fuels, drivers, and research

scientists (Caniëls and Romijn 2008). Another key issue is whether the potential economic advantages of biofuels can be enjoyed while ensuring that biofuel production does not damage food security (Jumbe, Msiska and Madjera 2009).

4. See for instance "Status of Codes and Standards to Enable a Hydrogen Economy," HTAC slide presentation, July 2007, at ⟨http://www.hydrogen.energy.gov/pdfs/htacjuly07_status_codes.pdf⟩; accessed August 31, 2011.
5. The ENGVA is now known as NGVA Europe.
6. Keay-Bright (2000) has provided an excellent analysis of these EU-negotiated arrangements based on extensive review of the relevant documentation and interviews with a great many participants and this has been drawn upon extensively in this section.
7. As she notes, "There is no doubt that the establishment of the UNFCCC Kyoto Protocol at COP3 in December 1997 was a major driving force which helped to bring the negotiations to their final conclusion in July 1998" (Keay-Bright 2000: 25).
8. This publication also quotes the leaked document which was quoted originally in a *Globe and Mail* report.
9. Oran Young has convincingly argued that many international regimes can make valuable contributions even without enforcement capacities. These include "programmatic" regimes that develop and execute joint projects and "generative" regimes that lead to new understandings (Young 1999: 79–107). The SMP fits with these categories.
10. A short listing of just a few of these private voluntary programs provides a sense of their diversity: the Association Internationale des Reparateurs en Carrosserie; the Automotive Aftermarket Industry Association; the Automotive Component Manufacturers Association of India; the Automotive Industries Association of Canada; the Automotive Recyclers Association; the Brazilian Sugarcane Industry Association; the Canadian Petroleum Products Institute; the Caribbean Basin Ethanol Producers Association; the European Association of Automotive Suppliers (CLEPA); the International Automotive Task Force; the National Association of Automobile Manufacturers of South Africa; the Renewable Fuels Association; the Japan Automobile Standards Internationalization Center; the Japan Automobile Transport Technology Association; and the World Fuel Cell Council. As noted above, OICA includes 41 national industry associations.

REFERENCES

Akinci, B., P. G. Kassebaum, J. V. Fitch, and R. W. Thompson. 2008. "The Role of Bio-Fuels in Satisfying US Transportation Fuel Demands." *Energy Policy* 36: 3485–3491.

An, F. 2006. *The Chinese Fuel Economy Standards for Passenger Vehicles: How it Works, the Targets, the Effects*. Presentation to the European–Asian Policy Workshop on Cars, Climate and Energy, Berlin Germany, July 19, 2006. ⟨http://www.autoproject.org.cn/english/new_advance_en/Chinese%20FE%20Standards%20for%20Passenger%20Vehicles.pdf⟩; accessed March 2, 2009.

An, F., and A. Sauer. 2004. *Comparison of Passenger Vehicle Fuel Economy and Greenhouse Gas Emission Standards Around the World.* Arlington VA: Pew Center on Global Climate Change.

The Australian. 2009. "Barack Obama Toughens US Car Emissions Standards." May 19.

BBC. 2009. *EU Climate Package Explained*. BBC News, November 26, 2009. ⟨http://news.bbc.co.uk/2/hi/europe/7765094.stm⟩; accessed January 11, 2010.

Bramley, M. 2002. *The Case for Kyoto: The Failure of Voluntary Corporate Action*. Prepared by the Pembina Institute in cooperation with the David Suzuki Foundation, October ⟨http://pubs.pembina.org/reports/VCR_publication_101702.pdf⟩; accessed April 16, 2010.

Caniëls, M. C. J., and H. A. Romijn. 2008. "Actor Networks in Strategic Niche Management: Insights from Social Network Theory." *Futures* 40: 613–629.

Corporate Europe Observatory. 2007. *Car Industry Flexes its Muscles, Commission Bows Down*. Briefing Paper March 16. ⟨http://www.corporateeurope.org/carlobby.html⟩; accessed October 31, 2008.

Demerse, C., and M. Bramley. 2008. *Choosing Greenhouse Gas Emission Reduction Policies in Canada*. Pembina Institute: B41, ⟨http://pubs.pembina.org/reports/pembina-td-final.pdf⟩; accessed February 1, 2010.

ERA-NET Transport. 2005. *Benchmark of Practices of Transnational Cooperation in Transport Research Programs and Policy*. October 2005. ⟨http://www.transport-era.net/fileadmin/fe_documents/80/ENT_WP1_DEL_1_3_v20_Final.pdf⟩; accessed December 23, 2009.

ERTICO. 2005. *History*. ⟨http://www.ertico.com/en/about_ertico/history⟩; accessed February 8, 2005.

EUCAR and CLEPA. 2009. *R&D Priorities for the Greening of Vehicles and Road Transport*. May, available at ⟨http://www.eucar.be⟩; accessed December 23, 2009.

European Commission. 2007. *Proposal for a Regulation of the European Parliament and of the Council: Setting Emission Performance Standards for New Passenger Cars as Part of the Community's Integrated Approach to Reduce CO_2 Emissions from Light-Duty Vehicles*. Brussels, December 12, 2007.

European Natural Gas Vehicle Association. 2004. *Move to Worldwide Harmonization of NGV/H2 Vehicle Standards? ENGVA News*, January 2004. www.sorpa.is/bindata/Documents/FinalENGVANewsJanuary1.20041.00101.pdf; accessed 2005.

European Union. 2009. *Green Cars*. ⟨http://ec.europa.eu/research/industrial_technologies/lists/green-cars_en.html⟩; accessed December 23, 2009.

Featherstone, M., N. Thrift, and J. Urry, eds. 2005. *Automobilities*. Thousand Oaks CA: Sage.

FIA Foundation. 2010. ⟨http://www.fiafoundation.org/about/Pages/AboutHome.aspx⟩; accessed February 1, 2010.

Finnemore, M., and K. Sikkink. 1998. "International Norm Dynamics and Political Change." *International Organization* 52 (4): 888–917.

Global Fuel Economy Initiative (GFEI). 2009. *50 by 50 Report*. March 17, 2009. ⟨http://www.50by50campaign.org/publications/Pages/PublicationsHome.aspx⟩; accessed February 1, 2010.

Global Fuel Economy Initiative (GFEI). 2010. ⟨http://www.50by50campaign.org⟩; accessed February 1, 2010.

Government of Canada. 2007. *Motor vehicle Fuel Consumption Standards Act Proclaimed*. Transport Canada Press Release No. H 215/07, November 7. ⟨http://

www.tc.gc.ca/eng/mediaroom/releases-nat-2007-07-h215e-2339.htm⟩; accessed February 1, 2010.

Hall, P. A. 1993. "Policy Paradigms, Social Learning, and the State: The Case of Economic Policymaking in Britain". *Comparative Politics* 25 (3): 275–296.

Hammerschlag, R., and P. Mazza. 2005. "Questioning Hydrogen." *Energy Policy* 33: 2039–2043.

Hira, A., and L. Guilherme de Oliviera. 2009. "No Substitute for Oil? How Brazil Developed its Ethanol Industry." *Energy Policy* 37: 2450–2456.

IEA. 2004. *Biofuels for Transport: An International Perspective*. Paris: OECD.

IEA. 2009. *Transport, Energy and CO_2: Moving Toward Sustainability*. Paris: IEA/OECD.

Institute for European Environmental Policy. 2003. *Service contract to carry out economic analysis and business impact assessment of CO_2 emissions reduction measures in the automotive sector*. By Patrick ten Brink, Ian Skinner, Malcolm Fergusson and Dawn Haines, Brussels.

International Council on Clean Transportation (ICCT). 2007. *Passenger Vehicle Greenhouse Gas and Fuel Economy Standard: A Global Update*.

Jaccard, M., and C. Bataille. 2002. "If Sustainability is Expensive, What Roles for Government and Business? A Case Study of GHG Reduction Policy in Canada." *Journal of Business Administration and Policy Analysis* 30–31: 149–184.

JAMA. 2009a. *On the Road to Sustainable Mobility: Achieving Greater Safety and Environmental Protection in Road Transport*. Japan Automobile Manufacturers Association.

JAMA. 2009b. *2009 Report on Environmental Protection Efforts: Promoting Sustainability in Road Transport in Japan*. Japan Automobile Manufacturers Association.

Jumbe, C. B. L., F. B. M. Msiska, and M. Madjera. 2009. "Biofuels Development in Sub-Saharan Africa: Are the Policies Conducive?" *Energy Policy* 37: 4980–4986.

Keay-Bright, S. 2000. *A Critical Analysis of the Voluntary Fuel Economy Agreement, Established between the European Automobile Manufacturers and the European Commission, with Regard for its Capacity to Protect the Environment*. European Environmental Bureau, Document No. 2000/021. ⟨http://www.eeb.org/publication/2001/ACEA-10-final-complete.pdf⟩; accessed November 17, 2007.

Kutas, G., C. Lindberg, and R. Steenblik. 2007. *Government Support for Ethanol and Biodiesel in the European Union*. Global Subsidies Initiative, International Institute for Sustainable Development, October 2007.

LoBianco, T. 2009. "New Policy Merges Fuel Efficiency, Tailpipe Limits; Rules Would Impact Design of Future Vehicles." *Washington Times*, May 19: A1.

Lucas, A., and V. Potes. 2006. *Voluntary Greenhouse Gas Emission Reduction: The Rise and Fall of VCR Inc*. Paper No. 2 of the Alberta Energy Futures Project, Institute for Sustainable Energy, Environment and Economy. ⟨http://www.iseee.ca/files/iseee/ABEnergyFutures-02.pdf⟩; accessed February 1, 2010.

Lutsey, N., and D. Sperling. 2007. "Canada's Voluntary Agreement on Vehicle Greenhouse Gas Emissions: When the Details Matter." *Transportation Research Part D* 12: 474–487.

Malakoff, D., and R. F. Service. 2002. "Bush Trades Hybrid for Hydrogen Model." *Science* 295 (January 18): 426–427.

Michaelis, P., and P. Zerle. 2006. "From ACEA's Voluntary Agreement to an Emission Trading Scheme for New Passenger Cars." *Environmental Planning and Management* 49 (3): 435–453.

Mikler, J. 2005. "Institutional Reasons for the Effect of Environmental Regulations: Passenger Car CO_2 Emissions in the European Union, United States and Japan." *Global Society* 19 (4): 409–444.

Mikler, J. 2007. "Varieties of Capitalism and the Auto Industry's Environmental Initiatives: National Institutional Explanations for Firms' Motivations." *Business and Politics* 9 (1): Article 4.

Mikler, J. 2009. *Greening the Car Industry: Varieties of Capitalism and Climate Change.* Cheltenham UK and Northampton MA: Edward Elgar.

Miller, S., B. Bahree, and J. Ball. 2002. "A Global Journal Report: Europe Launches Hydrogen Initiative." *Wall Street Journal*, October 16.

Mittelstaedy, M. 2002. "Firms Breaking Promises on Emissions, Study Says." *Globe and Mail* October 18.

National Research Council. 2008. *Review of the Research Program of the FreedomCAR and Fuel Partnership: Second Report, Committee on Review of the FreedomCAR and Fuel Research Program, Phase 2: 110.* ⟨http://www.nap.edu/catalog/12113.html⟩; accessed December 22, 2009.

New York Times. 2009. "Getting Ethanol Right." Editorial, May 24: 7.

OICA. 2006. *Worldwide Fuel Charter*. Fourth Edition, September: i. ⟨http://oica.net/wp-content/uploads/2007/06/wwfc-fourth-edition-sep-2006.pdf⟩; accessed February 1, 2010.

OICA. 2010. *CO_2* [brochure], p.7. ⟨http://oica.net/wp-content/uploads/climate-change-and-co2-brochure.pdf⟩; accessed February 1, 2010.

Oil & Gas Journal. 2009. "Politics and Fuel Choice" Editorial. August 10.

Onoda, T. 2008. "Review of International Policies for Vehicle Fuel Efficiency." IEA Information Paper in Support of G8 Plan of Action. Paris: IEA.

Paterson, M. 2007. *Automobile Politics: Ecology and Cultural Political Economy.* Cambridge: Cambridge University Press.

Piebalgs, A. 2009. "The Future of Biofuels." eBio, 1st European Bioethanol Fuel Conference, Brussels, April 1. ⟨http://europa.eu/rapid/pressReleasesAction.do?reference=SPEECH/09/171&format=HTML&aged=0&language=EN&guiLanguage=en⟩; accessed January 11, 2010.

Porter, I. 2009. "Obama Accelerates US Emissions Rules." *Go Auto*, May 20. ⟨http://www.goauto.com.au⟩; accessed June 9, 2009.

Porter, T. 2002. *Technology, Governance and Political Conflict in International Industries*. London: Routledge.

Pryor, F. L. 2009. "The Economics of Gasahol." *Contemporary Economic Policy* 27 (4): 523–537.

Reuters. 2009. "Toyota to Ask New Japan Government to Extend Green Car Aid." Sepember 2. ⟨http://www.reuters.com/article/idUSTRE5811YQ20090902⟩; accessed January 11, 2010.

Revkin, A. C. 2009. "Industry Ignored its Scientists on Climate." *New York Times*, April 24.

Rohde, P. 2005. "Rift in DOE Fuel Cell Program Exposed During Hydrogen Conference." *EnergyWashington Week* 2(14), from LexisNexis.

SAP. 2011. "Compliance Assessment for Sustainability Regulations in the Automotive Industry". Solution Brief, ⟨http://download.sap.com/download.epd%3Fcontext%⟩; accessed August 31, 2011.

Sauer, A., P. Mettler, F. Wellington, and G. G. Hartmann. 2005. *Transparency Issues with the ACEA Agreement: Are Investors Driving Blindly?* SAM Group/World Resources Institute, ⟨http://www.sam-group.com/downloads/studies/ACEA_Driving_Blindly.pdf⟩; accessed November 17, 2006.

Shiroyama, H. 2007. "The Harmonization of Automobile Environmental Standards Between Japan, the United States, and Europe: The 'Depoliticizing Strategy' by Industry and the Dynamics Between Firms and Governments in a Transnational Context." *Pacific Review* 20 (3): 351–370.

State of California. 2009. ⟨http://www.hydrogenhighway.ca.gov/about/about.htm⟩; accessed December 24, 2009.

Stigson, B. 2004. "Foreword." In Sustainable Mobility Project (SMP). *Mobility 2030: Meeting the Challenges to Sustainability*. World Business Council for Sustainable Development.

Sustainable Mobility Project (SMP). 2001. *World Mobility at the End of the Twentieth Century and its Sustainability*. WBCSD.

Sustainable Mobility Project (SMP). 2004. *Mobility 2030: Meeting the Challenges to Sustainability.* World Council for Sustainable Development.

UNEP. 2010. ⟨http://www.unep.org/pcfv/news/news.asp⟩; accessed February 1, 2010.

US DOE. 2009a. *Advancing New Vehicle Technologies and Fuels*. Fact Sheet, US Department of Energy. ⟨http://www1.eere.energy.gov/vehiclesandfuels/pdfs/vehicles_fs.pdf⟩; accessed November 24, 2009.

US DOE. 2009b. *Clean Cities*. Fact Sheet. US Department of Energy, November 2009. ⟨http://www1.eere.energy.gov/cleancities/pdfs/46689.pdf⟩; accessed November 24, 2009.

US Government Accountability Office. 2008. "Hydrogen Fuel Initiative" January.

VCR. 2002. "Voluntary Reduction Agreements May Shape GHG Regulation." *Champion News* 5(2), June. ⟨http://www.vcr-mvr.ca/June02_e_3.cfm⟩; accessed August 13, 2002.

VCR. 2005a. *When is it happening*. ⟨http://www.vcr-mvr.ca/vcr_when_e.cfm⟩; accessed January 1, 2005.

VCR. 2005b. *Frequently Asked Questions*. ⟨http://www.vcr-mvr.ca/assets/pupun/vcr_faq_e.cfm⟩; accessed January 2, 2005.

Westcott, P. C. 2009. "Full Throttle US Ethanol Expansion Faces Challenges Down the Road." *Amber Waves* 7 (3): 28–35 (figures on p. 29).

Winter, Drew. 2009. "Yes We Can, Maybe." *Ward's Auto World*, June 1: 2.

World Forum for Harmonization of Vehicle Regulations (World Forum). 2004. *Road Map to Develop Hydrogen/Fuel Cell GTR(s)*. Informal document GRPE-48-17, June 4.

WWFC Committee. 2009. *Ethanol Guidelines*. ⟨http://www.cvma.ca/eng/publications/E100_Guideline_final_26Mar09.pdf⟩; accessed February 1, 2010, and WWFC Committee. 2009. "Biodiesel Guidelines." ⟨http://www.cvma.ca/eng/publications/B100_Guideline_final_26Mar09.pdf⟩

Yacobucci, B. D. 2008. "Hydrogen and Fuel Cell Vehicle R&D: FreedomCAR and the President's Hydrogen Fuel Initiative." Congressional Research Service Report for Congress, updated March 20.

Young, Oran R. 1999. *Governance in World Affairs*. Ithaca NY: Cornell University Press.

Part IV

Conclusions

7

Private voluntary programs in climate policy: Potentials and pitfalls

Karsten Ronit

Introduction

Climate issues have come to the fore of global politics, with private voluntary programs occupying an intriguing, yet unsettled, role. Indeed, these schemes are unfolding at a time when the rules guiding climate policy and its institutions are in the critical stages of being established and reformed at the intergovernmental level, and when transformations are taking place in the business community, slowly and hesitantly giving more substance to climate-friendly strategies. Alongside these processes an array of civil society initiatives highlights the importance of halting climate change both in big politics and in the everyday life of citizens. In addition, assessments and predictions from a cross-disciplinary landscape of scientists are pointing to the urgency of timely strategies and the need for new and robust mechanisms to monitor rules and enhance compliance.

Nonetheless, the recent story of climate change politics and private voluntary programs not only reports the achievements that have been made; it also shows the failures that are evident, and where optimism remains unchecked. Indeed, if effective private voluntary programs are to emerge it will take time for them to materialize and consolidate. We are in an experimental stage of governance innovation, and we cannot make any final judgments about the prospects for private voluntary programs. We can, however, point to strengths and weaknesses and offer some guidance for their future development.

Business and climate policy: The potentials and pitfalls of private voluntary programs, Ronit (ed.), United Nations University Press, 2012, ISBN 978-92-808-1214-5

In this book we have referred to *private voluntary programs*; and what was stressed in the introductory chapter, namely, that this is a complex entity, can now be further qualified. Although private voluntary programs share a number of common features in climate policy, they come in many forms and are given different names and theoretical labels. Hence to accommodate these variations we distinguish between different degrees of “privateness”, “voluntariness” and “programness”.

As to the private character of programs, we need to differentiate between various degrees of “privateness”. The aim of private regulation is to influence various types of business behavior that are not already laid down in existing corporate practices. From this, however, we should not conclude either that programs become fully public when public institutions sponsor them, or that programs become completely private in cases when action is launched by business, perhaps assisted by other private actors. We have also seen that business authority is not kept entirely undisturbed when programs are generated by business, and that authority is not completely surrendered in those cases when public institutions draft them. Thus, “privateness” is prevalent in the hybrid forms of programs, but, of course, they must entail a significant private component to separate them from public regulation.

Furthermore, we must distinguish between different degrees of “voluntariness”. The choice of creating and joining a program, and continuing to support it, is informed by many contextual factors that enable or constrain decisions. Firms and industries are often very conscious of such factors. This, however, should neither lead us to the conclusion that firms and industries act only under strong pressure and exclusively from a defensive position, nor force us to attribute the concept of private voluntary programs only to those schemes that are supported without any kind of identifiable threat, or even encouragement, from forces outside business. Altogether, “voluntariness” comes in many shades.

Finally, it is helpful to distinguish between different degrees of “programness”. Programs differ widely in their formalization, in their codification of rules, and in their organizational structures. From this, however, we should neither reserve the concept of program only for those schemes that have already reached a high degree of rule elaboration and are endowed with significant resources to run these programs, nor limit ourselves to employing the concept to characterize the looser arrangements as programs and reserve other concepts for the more established ones. Rather, “programness” can be ranked along a continuum, enabling us to analyze and evaluate a variety of programs.

As the contributions in this book demonstrate, it is useful to accommodate climate change schemes under the overall concept of *private voluntary programs*. Indeed, “voluntariness”, “privateness” and “programness” are generic features of any program; therefore many different concepts of private regulation, also discussed in this book, grapple with these same essential issues. Having discussed the scope of these different programs, we can now explore in further detail the many experiences revealed by the studies in this book.

Public policy, private goals

The relationship between public policy and private goals entails several interesting dimensions. Is it possible to mitigate corporate interests with those of society? Can a diversity of future goals become integrated into current programs? And is coordination of a multitude of programs achievable? The chapters in this book illustrate these problems in different ways, and there is not always a perfect misfit or even a comfortable fit. Public policy can be difficult to marry with the private goals of business. It is notable, however, that over time, the roles of the public and private sectors are also changing, a feature observed in many other areas of public policy (Hirschman 1982).

First, many private voluntary programs emphasize the perils of climate change as the primary background for embarking on a new course, and observe the public good dimension when launching initiatives. Existing programs, typically with an environmental focus or an ambition to save energy, integrate climate concerns, while new programs often depart from a more distinct climate agenda. Thus, environmental and energy policies develop into climate policy.

Programs highlight climate issues in various ways. They may refer to public debates, to scientific reports, and to the negotiations of governments, and they may highlight specific challenges, such as the burdens transferred to future generations or the consequences for different welfare parameters. But they also point to the possible losses to be suffered or the potential benefits to be gained for business in general, for particular industries, or for individual firms when adapting to post-carbon production. Therefore, goals adopted from various political and scientific contexts are not simply reproduced; they are blended with private agendas and adjusted to specific corporate opportunities, and, consequently, they are not straightforwardly amenable to the smooth implementation of public goals. This problem is most evident in cases where programs emerge in the private sector as shown by Fuchs and Boll (Chapter 5), and is less pronounced in cases when public institutions draft them, as seen in the publicly initiated programs in the US dealt with by Hsueh and Prakash (Chapter 3), or in cases where other private parties beyond business have been active in creating them, such as in forestry as analyzed by Bozzi, Cashore, Levin and McDermott (Chapter 4).

In areas such as renewables, where business through the non-emission of greenhouse gases contributes to halting climate change by virtue of expanded business activity, it is relatively easy to unify private and public goals. Indeed, solutions here are not necessarily guided or underpinned by private schemes but more simply provided through the introduction of new technologies. In areas in which a reduction of emissions is tabled, a more delicate weaving together of goals must take place, and in areas struggling with high emissions, significant problems may arise in the process of aligning goals, as shown by Porter in his study on automobiles (Chapter 6). Invoking new business strategies is also on the agenda in those parts of business that by way of their transboundary role in the economy, such as the financial industries, are linked to established industries and can redirect investments to more climate-friendly sectors. As illustrated by Clapp and Thistlethwaite (Chapter 2), new institutions are created to that end, but they are still underdeveloped and characterized by enforcement problems.

It is generally recognized that swift answers to these issues are not in sight. Despite various conflicts over goals, it is notable that the political atmosphere in which many programs are conceived is favorable to incorporating public goals, and that programs communicate some of the societal orientations on which they are founded. Although not all communication, of course, should be taken at face value, it is nevertheless interesting to observe that programs are, in part, clad in such a language.

Second, adding significantly to the complexity of goal formation, is the time horizon of corporations. In principle, private voluntary programs must enhance the economic performance of firms, or at least confirm the status quo in the marketplace, so that participants are not losing ground to competitors. If initiatives are not instantly successful, future gains must still be possible to identify, and, therefore, drawbacks must not last too long and lead to the decline of business activity, ultimately threatening survival. This is a particular challenge for private voluntary programs. Indeed, firms may perish in a heroic effort if initiatives are prohibitively expensive or cannot be sustained. It should be noted, therefore, that economic parameters are flexible to some extent – investors, producers, retailers as well as customers may give a high priority to the introduction of innovative technology or new sustainable products and services and make climate-friendly strategies more likely; but benevolent and patient action by the few is not enough, as scale matters enormously in halting climate change.

Out of grim necessity, immediate action is required in climate change politics in order to make the most of precious time, although a number of climate goals can only be reached in the more distant future. Private voluntary programs, therefore, face serious dilemmas. For one thing, business is focused on immediate gains to continue production and attract investments; for another, business must operate in a regulatory environment in which predictability is crucial, and in which investments are only amortized over the long term, and with new technologies risk may be particularly high. This is very evident in Porter's analysis of the automobile industry, where developing alternatives to current technologies and their infrastructures is dauntingly complex and uncertain. Consequently, time is a key variable that affects the status of private voluntary programs in the overall governance architecture.

The impact of private voluntary programs on greenhouse gas emissions and the natural world will take time to evolve, complicating an evaluation of their effectiveness. Although accounting standards have been established by private actors such as the World Resources Institute (WRI) and the World Business Council for Sustainable Development (WBCSD), and the Intergovernmental Panel on Climate Change (IPCC) and the International Organization for Standardization (ISO) are also involved, greenhouse gas accounting is very complicated. As shown across the chapters, many measures aim more at cost saving than at "climate saving". In this context, it must be emphasized that the creation and stability of programs are not necessarily enough to bring about the changes needed. Indeed, governance effectiveness may lead to a false sense of security because there is scientific uncertainty about what is really needed to halt climate change, and how these demands may change over time.

However, firms can be assisted in various ways to compensate for some of the inherent problems related to the time factor in private schemes. States have a long time horizon, and although they may also face decline relative to other states, they are not confronted with the same type of risks as business. Regulation favoring sustainability can be adopted, and states and intergovernmental organizations can help private voluntary programs by sponsoring them, or by supporting them once they are established. Such measures will have an impact on the time horizon and can help solve some of the paradoxes in business between immediate calculations and future concerns. Bozzi, Cashore, Levin and McDermott's chapter also, however, points to the reverse arrangement between public policy and private goals, in which the "symbiotic" relationship between the two bolsters public policy. One benefit of layering on a voluntary private program to fix problems in the public policy process is that this bypasses hurdles inherent in reopening the negotiating process and possibly re-engaging in hard-won battles.

Third, based on a significant amount of creativity, many climate-related programs have actually come into being. As shown by Fuchs and Boll, firms in the food sector have alone or with other firms adopted some programs, but fragmentation persists, while Bozzi, Cashore, Levin and McDermott show that private regulation has become more centralized, although there is institutional overlap and some initiatives refer to each other. Porter points to the variety of ways that programs come into being, ranging from those responding directly to initiatives of public authorities to those initiated more independently by business. In addition to these various sector-related programs, Hsueh and Prakash analyze how the US government promotes several programs that firms from different sectors can join, and Clapp and Thistlethwaite scrutinize a variety of new schemes and entities in the financial industries that develop programs of relevance to business at large. Beyond the cases studied in this book, we find many programs that raise awareness, formulate rules, and sometimes also police them, making the total proliferation of programs quite significant.

However, the emergence of private voluntary programs presents two different narratives: there are too many programs, and there are too few programs. We have observed an oversupply of programs; there are many and occasionally overlapping initiatives involving different groups of firms, embracing different industries, specifying different commitments, communicated by different actors and located at different territorial levels. At the same time, we witness an undersupply of programs; in many areas of business activity, programs have not emerged, or their goals are not sufficiently clear, not all relevant firms are included, tools to implement them are weak, and they are not yet suitable to complement or replace public regulation.

This institutional pluralism is attributable to the many and uncoordinated business initiatives and also to the lack of guidance on the part of civil society, government and intergovernmental organizations; their encouragement is directed not so much at unifying programs as at stimulating initiatives more basically. Today it is barely possible to inventory the many and diverse programs currently in place, and neither private nor public agencies record them. A key factor behind this kind of pluralism is, no doubt, that climate change politics and its various governance structures are still immature, and only at a later stage in program development can greater coherence be envisaged. These boundaries to coordination should be analyzed in a broader context, however. Given the quite extraordinary scope and complexity of climate change, it is not possible simply to extrapolate from other fields of public policy, and there seems to be a high demand for different regulatory entities both within and across industries, a point that will be discussed further in the next sections.

Institutional conditions: The changing roles of key actors

The development of private voluntary programs in climate policy poses several questions concerning the content of goals, the time horizon of strategy, and the multiplicity of programs. How the public and private dimensions are combined is inexorably a challenge in the study of emerging governance structures (Delmas and Young 2009) and in the analysis of regulation (Gunningham 2007). We therefore have to examine the key categories of actors involved in establishing and running programs if we are to fully appreciate the potentials and pitfalls of these schemes.

First, business is a decisive force behind private voluntary programs – even if other public or private actors create them it is crucial that business supports and implements them. This pivotal point does not exclude external factors, of course, and thus the capacities hosted and employed by business are properly understood only against the backdrop of such factors. This book has taken us beyond some of the firm-centered analysis found in the traditions of corporate social responsibility and business ethics studies, and we need analyses that give greater weight to political aspects (Vogel 2008).

Private voluntary programs are in some cases closely linked with traditional public policy, with business interacting with governments and with intergovernmental organizations. This is, of course, most distinct in cases in which public agencies offer multisectoral programs for firms to sign up to, as demonstrated by Hsueh and Prakash. The public background for private action is important but the irony is, however, that in this case it seems to be a weak rather than a strong state that engages in these sponsoring initiatives.

Business, however, also responds to other public strategies, including those that prompt preemptive action on the side of business and lead to private initiatives. Nevertheless, it is vital not to exaggerate public direction. Only occasionally do we see specific industries explicitly referring to goals adopted in a public framework – beyond the general ambitions in climate policy, of course. Exceptions include the role played by the auto industry in the European voluntary programs, which responded to specific emissions goals negotiated with public authorities, or the role of the auto industry in public–private collaborations intended to invest in new technologies.

It would also be difficult for business to formulate clear ambitions, given the relatively limited operationalization of climate goals valid for particular industries and made available through states and intergovernmental organizations. As a matter of fact, we find evidence of mutual disregard, and a number of private voluntary programs exist more or less in isolation from public institutions because they are mainly driven by the internal dynamics of business. Developments in the food industry, studied by Fuchs and Boll, demonstrate how these internal processes work and that the degree of program activity does not vary systematically with the extent of public involvement.

Programs are located at very different levels in the business community. Single firms, most typically large corporations, communicate a climate message. However, these initiatives are not legally binding and they are difficult for other actors to monitor, especially if they are not tied to other, more encompassing industry programs. This requires, among other things, that active and critical publics keep watch over the commitments of individual firms, which is a huge task, given the many corporate activities. Measured against the otherwise limited resources of relevant independent monitors, many such initiatives are likely to remain relatively unobserved, as we have seen in the food sector.

Substantial programs are being built at the level of industries, and the contributions of this book stress that single-firm activities are embedded in different forms of collective action. Indeed, it would be strange if established business interest associations, that had historically developed a capacity to cater to a variety of economic and political preferences of the industries they represent, completely ignored the emerging field of climate policy. Over time, business interest associations have entered new domains and taken on new responsibilities, and thus a task expansion would in no way be new to them. Not everywhere, however, do associations play a strong role in the life of private voluntary programs, and various organizational dynamics lead to the design of other types of outfit. In the case of the automobile industry, its oligopolistic structure, and the need to bring in entirely new activities, such as the development of biofuels, has contributed to the weakness of traditional industry associations and the growth of new ones. At the global level there are relatively few trans-industry associations to facilitate collective action (Porter and Ronit 2010) and, therefore, the emergence of new, broader, climate-specific entities often takes place outside established frameworks (Meckling 2011).

Quite interestingly, new entities emerge in business, either in the form of new and specialized organizations, such as the International Emissions Trading Association (IETA), or brand-new alliances expressed through initiatives, such as the Copenhagen Statement on Green IT in 2009 initiated by Microsoft, Konica Minolta, Intel, HP, Fujitsu, Dell and Cisco. In some cases, established organizations may simply be too conservative, refrain from engaging too strongly in climate issues, and avoid committing their members. As an example, several large corporations decided to leave the American Chamber of Commerce in fall 2009 because of its limited engagement in climate issues.

There are also several examples of existing associations embracing some aspects of climate change in the different areas covered in this book, but also new organizations are in demand when prompt action is needed, when traditional associations are split, or when challenges in various ways cut across existing forms of organizing business interests and demand coordination. Evidence of such a pattern is brought by the study of Clapp and Thistlethwaite on the financial industries and their novel organizations, which offer a particularly good illustration of this. If new initiatives, however, can only draw on a smaller membership base, and if entirely new resources must be harnessed to run programs without support from existing structures, there are also barriers and disadvantages in forming climate-specific entities.

In such cases parts of the organizational capital in and across industries risk getting lost. Thus, a major challenge is to draw on existing entities to respond to new demands in the membership and in society and only to create new entities where appropriate. Further, it is important that new bodies do not bring confusion by breathing life into a huge number of initiatives which blur the accountability of business.

Second, states and intergovernmental organizations have in some areas been influential in setting agendas in climate policy, sometimes with the intent to deny climate change and block initiatives, and sometimes with the ambition to formulate strategies and implement rules more generally. In spite of serious difficulties and impasses a certain momentum has been reached, however, especially after the Obama Administration took office, and greater agreement has been achieved in recent years. It should be borne in mind, however, that governments and regional players like the EU have not just been waiting for an international breakthrough but have invigorated programs at these levels. States and regions, again with significant variations, pay attention to the various opportunities of private initiatives, whereas global public policy is not well-coordinated across the many public agencies in and beyond the UN system.

Beyond the areas documented in this volume, many other cases of intergovernmental agencies exchanging with particular industries can be found, and these experiences also set our findings in perspective. Shipping and airlines, for instance, were exempt from the Kyoto Protocol, but in many ways specialized bodies within these fields have proved to be relevant not only in setting more concrete agendas but also by playing an active part in rule-making. In this context, it is interesting to note that established business interest associations such as the International Chamber of Shipping (ICS) have had a significant input into negotiations with the International Maritime Organization (IMO) and the development of rule-making to reduce emissions (ICS 2010). In addition, a major private initiative has been developed in the airlines sector through the

International Air Transport Association (IATA) without waiting for directive action from the International Civil Aviation Organization (ICAO) as the designated body for air traffic, but calling for joint cooperation with this special UN agency as well as with civil society stakeholders (IATA 2009). It is interesting to see that other special UN agencies and key business interest associations have increasingly sought to coordinate efforts and reach beyond the more general messages of states and intergovernmental organizations. The World Tourism Organization (UNWTO), for instance, in close cooperation with the World Travel and Tourism Council (WTTC), has assisted in giving this business interest association a stronger role in the travel and tourism sector (UNWTO and WTTC 2009).

The role of UN agencies varies, and each industry does not have its "own" specialized counterpart in the intergovernmental system, so such patterns of exchange are not generally found across business. However, as the chapter by Bozzi, Cashore, Levin and McDermott shows, we have seen general encouragements in the context of the Conferences of the Parties (COPs) to private arrangements in forestry and the ecosystem surrounding it. Some further development of the governance architecture might, however, find ways to cope with the challenges of policy coordination, and major steps have now been taken through the REDD+ initiative.

We therefore find different types of public policy frameworks – national, regional and global – behind private voluntary programs. However, links are very unevenly developed. A public "shadow" can be found along a continuum from inaction to action, with the latter including cases from broad encouragement to targeted interventions, but currently private voluntary programs are not sufficiently integrated into an overall strategy of climate change politics. It might be expected, however, that public authority would have some unifying impact on an otherwise relatively fragmented business community, although the experiences from this book's contributions indicate that such centripetal forces are hard to identify more generally.

In cases when private actors take action, a generally welcoming attitude from public authority seems more common than an express provision of assistance to particular programs. Direct stimulation is missing in relation to the food industries, as described by Fuchs and Boll; and in the case of the financial industries, as discussed by Clapp and Thistlethwaite, support is negligible. As examined by Porter, it is also quite revealing that public policy tends rather to unfold at national and regional levels, although several agencies take a specific interest in transportation issues. The lead at the intergovernmental level has been taken by the International Energy Agency (IEA) in its work on transportation more generally. Some pressures can be observed from various intergovernmental agencies but they are generally more diffuse, although solutions here might contribute to halting climate change in an area that is otherwise known for having a sizeable share in emissions.

Third, the growing importance of climate policy is attributable to the active role of many civil society groups, whose strategy is to influence business behavior through different channels. Typically, they leverage states and are present in the corridors of intergovernmental negotiations without formally being a part of them, and thus they have sought to influence business through traditional public policy. Major conflicts have characterized climate change politics, and civil society groups have mainly tried to engage in the big issue of whether climate change is actually human-caused. As long as this scientific, political and ideological battle went on, this was the main emphasis for these groups – as well as for most states and for expert groups. Hence, relatively limited attention was directed towards the issue of how to combine public and private efforts.

It follows from this that these groups have generally been stronger in the normative field of climate change politics and have had less capacity with regard to problem-solving in specific industries. From a strategic point of view, it is far easier to operate at the more general level where industry-specific knowledge is not demanded. It is also easier to support, for instance, initiatives in the renewables sector and to stress the need for alternatives than to enter into a dialog and build a program with key emitters and keep this afloat. Indeed, the sharing of responsibility can be risky and can damage reputation.

However, civil society groups have been involved in program activities indirectly and directly – and some efforts are related to their more general influence on traditional public policy. In cases analyzed in this book – where governments, such as in the US, documented by Hsueh and Prakash, or regional cooperation of governments, such as the EU, emphasized by Porter, have more specifically created programs for firms to join – these activities can be seen, to some extent, as resulting from the inputs delivered by civil society groups. Bozzi, Cashore, Levin and McDermott point to an additional, "symbiotic" example in which technical programs developed in civil society layer on top of public regulation and provide incentive for businesses to adapt their behavior to incorporate co-benefits.

Although major parts of implementation in publicly sponsored programs rest with business, regulation is not completely surrendered to corporate players. There is certainly variation in the degree to which civic groups engage with business around programs. General hesitance towards the whole idea of private voluntary programs in their more distinct private versions is definitely one of the barriers to civic groups playing a more active role. As mentioned already, resources are limited on the side of civic groups, and this is likely to confirm a tradition of more general debate rather than practical cooperation with business, but in some cases civil society groups assist in the organization of business interests.

It is also exciting to observe the variation across cases. It is not very surprising that civic groups are poorly integrated in the arrangements of the financial industries, as emphasized by Clapp and Thistlethwaite, although certain initiatives have a strong civil society component. We know from other studies of the financial industries and the intergovernmental organizations in this field that there is a tradition of excluding these groups (Scholte 2002). This is hardly conducive to building up any civil society tradition or competence as far as private regulation is concerned, although the new climate agendas of these bodies and the new initiatives may change these patterns. The fairly limited contact between civic groups and business is in many ways more surprising in food and automobiles because these industries very much depend on the end-consumers. It seems, however, that consumers are appearing as individual actors in the market rather than as coordinated groups, with a potential to work in the context of programs more specifically, and consumer preferences may also work against climate-friendly initiatives.

The orientation of these civic forces towards climate issues does not rest on solid traditions, and is likely to be led by mixed motives. Some strategies favor launching critical campaigns and taking to the streets, while other strategies emphasize the need to become involved in a dialog. The account given by Bozzi, Cashore, Levin and McDermott for forestry is quite different from that about food and automobiles. Environmental groups have had a lead role in the formation and operation of a number of the voluntary forest carbon offset programs, such as the Climate, Community and Biodiversity Standards and the Voluntary Carbon Standard, illustrating more generally a stronger role for environmental organizations than for consumer groups and an opportunity to forge alliances with experts. Forestry is an exciting case, as the environment and climate are so closely connected that for the large and established environmental movement, it has been relatively easy to embrace this field.

Fourth, experts have been crucial in bringing climate change not just to the scientific agenda of a few specialized research communities but also to the global political agenda – in many cases other actors have followed in their train, including business. There are many skeptics among climate researchers; although there is broad agreement about human-caused climate change and how to avoid the special risk of tipping points in nature, many uncertainties remain about implementation. Based on scientific knowledge, experts have highlighted the importance of the time dimension: we need historical points of departure for measuring emissions, but we also need to fix certain targets in the future and establish a timetable for the gradual curtailment of emissions, although these targets are not always politically and economically possible or technically feasible. As emphasized by Clapp and Thistlethwaite, this is a particular challenge for programs in the financial sector because they relate to different industries. Accordingly, it is necessary to establish broader, recognized standards for measurement. Indeed, different actors interpret data differently, giving different priorities to them, and much of the discussion revolves around reliable and systematic knowledge.

Relevant and convincing data has become more available recently, and in many ways also more disquieting; therefore, business should be well-motivated, take action, and engage in the development of more and better programs. However, things remain complicated and, indeed, there is a lack of translation of evidence and coordination which could otherwise promote private voluntary programs, indicating that knowledge does not automatically spill over into private initiatives.

Experts are found across the organizational landscape of climate change politics, but many are in one way or another related to the public system and, as pointed out above, climate problems have to a large extent been interpreted and analyzed at a general level rather than at the level of particular industries. In essence, the major expert units in the study of climate are not focused on business or on specific industrial sectors but are focused on nature and its different subsystems. This is surely one of the reasons why expert knowledge has more easily been drawn into program-building in forestry – forests are conceived of as an ecosystem. Experts have, however, also been engaged in the food sector and automobiles, but much less so, and they seem to be much less involved in finding appropriate solutions for these industries. Other, significant dimensions of these industries, such as production characteristics, safety dimensions, ownership structures or organizational properties, are scrutinized, but expert knowledge is not predominantly linked to climate issues, and often environmental expertise must be brought together with engineering and other expertise. Combining different forms of knowledge is, therefore, an intriguing task.

Elements of schemes

Different actors bring different preferences and resources into private voluntary programs, and there are different ways to enhance their legitimacy and effectiveness; however, to gain a better grasp of the potentials and pitfalls of these programs, we need to identify further which kind of problems programs run into.

Ultimately, of course, the achievements and failures of private voluntary programs must be examined in relation to their effects on climate change. Institutional effectiveness of programs in terms of changing norms and behavior does not necessarily lead to the desired goals of halting climate change. Goals can be too weak, arrangements can come too late, and conditions in the natural environment can change in ways that are currently unpredictable and require a thorough reformulation of goals and a redesign of programs. The adequacy of policy instruments devised to halt change must be studied in a long-term perspective but, in general, programs are of fairly recent origin. Institutional potentials and pitfalls, therefore, more easily lend themselves to scrutiny in a short- or medium-term perspective, and in this book, institutional effectiveness is examined in relation to building private voluntary programs. The analysis includes core elements such as designing, joining, monitoring and complying – each of these elements showing different degrees of "voluntariness", "privateness" and "programness".

In fact, many things go wrong and several institutional imperfections can be listed: for example, programs cannot be designed, and they lack coordination with other relevant programs; firms do not join, and no ambitions are formulated to recruit them; nobody monitors business, and people rely instead on the declared intentions of firms; firms shirk, and rules and norms of programs are violated with no or limited consequences. However, this bleak picture does not stand alone. To some degree these failures are useful parts of the inevitable trial-and-error process that is needed. For instance, the problems of using hydrogen to fuel cars might not have been revealed without failed attempts to develop the technology. Moreover, several notable achievements with long-range potential have arisen in the emerging field of climate policy. Indeed, when assessing the potentials and pitfalls, it is necessary to bear in mind that the real world of private regulation should not be compared with the ideal world of public regulation – an error crippling a proper analysis. As a matter of fact, one of the reasons why private voluntary programs have emerged is that traditional public regulation has not been able to handle these issues in a timely and effective manner.

Designing

In some cases private voluntary programs are designed by public institutions, as we saw in the various multisectoral programs, highlighted by Hsueh and Prakash, and in some regions of automobile production, discussed by Porter; in principle, their design rests on a hierarchical decision and a separation between the regulator and the regulated. In this sense, the voluntary aspect is pushed into the background, but it is formulated as a particular method of regulation to be followed once the program is adopted. When programs emerge in a top-down fashion, the voluntary element consists of enrolling in a program rather than drafting it.

Such government-initiated programs may be steered by various political ambitions, and business is, therefore, not in control of the process in the same way as it would be if programs were designed by business alone or in cooperation with other private parties. In the cases of food, automobiles and the financial industry, programs have a strong business component. Here, however, voluntary action can be limited by another factor, namely, the ambition of the initiators of programs to include only some firms and exclude others from taking part in the creation of a program. Moreover, privately initiated voluntary action may be limited in terms of ambition due to a divergence between private interests and public objectives, as pointed out above and as shown by Fuchs and Boll. Thus, by adopting an inclusive or exclusive strategy in the process of designing, this will have an impact on the patterns of joining – as we will discuss shortly.

It is evident that the private component tends to be weak when programs are created by public institutions. That reality does not exclude, however, that some useful contacts have been made with business prior to the designing of programs, and this can be necessary to check possible implementation problems. However, business authority is also moderated or challenged when more parties are involved, and these other forces may bring other aspects into the design of a program, as they are more aware of how programs will be perceived beyond the business community. At least certification schemes such as the CCBA and Plan Vivo, as shown by Bozzi, Cashore, Levin and McDermott, are concerned about this and have a more advanced practice of transparency and communication; also, some novel arrangements in the financial industry cater to these problems. For example, many programs in the financial industry use disclosure to generate a market signal among investors and financial service providers that links a firm's contribution or exposure to climate change to financial risks.

Programs are not equally developed in terms of their organizational structures and the issues covered by the rules of the programs. Those designed by public agencies should have a solid backing; indeed, they must do eventually because it is easier for public institutions to observe the programs they have created for business, and intervene if they fail, than to police those designed by business and other parties. But it is clear that the "swords" employed in these programs are not always strong, as shown by Hsueh and Prakash. From a somewhat similar viewpoint, we should expect traditional business interest associations, or other established entities in the business world, to play a helpful role in rallying members around designing and reforming programs in ways that mitigate conflicts between member firms, but weaknesses are evident.

Joining

The abundance of programs tells us something about the successes of bringing them into being and the vigorous activity in climate change politics; however, it is essential that firms enroll to act for the benefit of the climate. Basically, joining is much easier if standards are low and commitments limited, while programs with stronger gatekeeper functions and privatizable benefits set limits to participation and exclude underperforming firms. Public agencies make programs available without resorting to traditional top-down regulation, but they cannot force firms to participate under the regime of private voluntary programs. In fact, the achievements of programs are in some ways reversed because government-designed programs, which are relatively easy to set up, are more difficult to put into force because they are not born from or through members. Hsueh and Prakash clearly show that government-initiated clubs with stringent standards are very challenging to those firms that join.

From a general perspective, programs that are initiated by business interest associations and other established organizations that already have a stock of members are usually born with a certain support, whereas new programs that specialize in setting standards in relation to climate change must carry out independent recruitment or conduct relevant climate change analyses to attract firms. If an association is involved in building a program, and if the members are compelled to join the program, firms become members by virtue of their affiliation to the association – not necessarily because they have an express preference for the program. There is, of course, a risk that firms lack motivation to support an initiative, but it can be a great advantage if representative associations with a negligible number of free-riders can mobilize their members. We do find evidence of traditional associations that address the challenges of climate change; however, in the cases of forestry, food and cars, and even finance, such existing platforms and mechanisms that could make participation more uncomplicated are still weakly developed. It is clear that a significant potential for developing private voluntary programs remains untapped.

In the automobiles case it is, nevertheless, easy to identify the producers and the potential members of a program. The industry is simply so concentrated that no firm can possibly be ignored, and hence, there is no broad and anonymous recruitment of car producers. But the conflict between continuing traditional production versus shifting to new vehicles with lower emissions is a big problem and sets clear limits on the creation and value of private programs. In the food industry also, we find concentration and a limited number of large firms – although to a far lesser extent than in the automobile industry – but strong programs with significant joining are missing. Affairs seem to be different in forestry, and there we have seen a strong effort to recruit new members to these programs. In some cases, however, advanced designs imply that those who join are strongly committed, but there are many firms that prefer not to join, and different programs exist.

There is a major difference between industries in the matter of joining; economic concentration and small groups are clearly key variables – a point we shall return to shortly. This is potentially quite promising if industry was to decide to run stronger programs and if single firms were to decide to abandon their own plans to the benefit of a collective effort; it can in many ways be relatively easy to rally a limited number of firms and draw upon the advantages associated with small group size, including spotting possible free-riders.

Monitoring

Private voluntary programs at all levels in business are often criticized for their superficial promises and hollow commitments that nobody really checks. Indeed, proper monitoring mechanisms are decisive in avoiding failure of existing programs and presenting them as a valid alternative to public regulation. The public-good character of avoiding climate change strongly emphasizes the public aspect of programs. Indeed, if it is possible to convince those not familiar with a particular scheme and having no inside information about the given industry that monitoring is taken seriously, then the monitored firms have much to gain. And if programs are working properly, they are more likely to attract new firms.

Two major problems are that the behavior of free-riders cannot always be appropriately examined and that firms cannot be made accountable to programs they do not subscribe to; with a good communication strategy, however, it is possible to correct information asymmetries and bring greater transparency into this field too, as emphasized both by Hsueh and Prakash and by Fuchs and Boll. It is possible to communicate who is inside and who is outside a given program so as to reserve major reputational goods for club members. This can be concluded, for instance, from the government-initiated programs, as well as from programs in finance and in forestry.

Monitoring members of a program is much easier, of course, than monitoring those standing outside. The programs examined in this book use different monitoring mechanisms, and these are closely linked to their administrative capacities. The fact that public institutions monitor the behavior of firms in government-initiated programs does not necessarily suggest that policing is very advanced, but it must be strong enough to deny and not extend membership to those firms that do not comply with the rules. Hsueh and Prakash offer illustrations of this. More advanced monitoring and verification is found in certain programs in finance and in forestry, and more specific procedures for monitoring and reporting are stipulated in these programs, focusing on some key performance criteria where firms must renew their membership on a regular basis. In the financial industry, as shown by Clapp and Thistlethwaite, monitoring is often initially designed using a "hierarchical" structure, but with the long-term goal of creating a "horizontal" structure where firms self-regulate and compete to reduce their exposure and contribution to climate change risks. This is of course more effective than having a firm become a member once and for all, or trusting a firm simply because of its membership. It is interesting to see that some programs are drawing much more on expert knowledge and the participation of other parties. They bring their own perspectives and capacities into programs and are able to qualify monitoring in ways that business alone rarely can.

However, other forms of monitoring are imaginable when there is no proper administration behind a scheme. Concentrated industries with a limited number of capacitated firms are better able to monitor the behavior of a few rivals, and they have strong economic incentives to monitor firms that violate standards. The Worldwide Fuel Charter, which is organized by auto associations, is an example of how an industry can work together to impose standards on suppliers, in this case of fuel, and to exclude non-compliant producers. Thus, selfish motives can be very helpful in contributing to solve public problems such as climate change. Key firms are also in a position to coordinate in the industry in more informal ways without necessarily activating some joint body. In such cases, however, it can be problematic if information on monitoring is not disclosed to the public, but different forms of formal and informal monitoring may complement each other.

Complying

Monitoring is meant to detect non-compliant behavior once it occurs, and also to deter firms from engaging in such activities. If a firm establishes its own program, compliance is an in-house affair, and the firm has to control its own performance. When firms join a government or privately initiated program, they must respect some rules together with other firms and refrain from shirking. This requirement brings them into quite a different control and enforcement situation. This handling of compliance issues shows that embedding programs in a broad institutional context is necessary and can be used more systematically to draw upon some of the existing resources in and around business.

There is a clear indication that those firms that join a program also comply. In a way, this may sound self-evident, because why should they join in the first place if not with the intention of keeping the rules? However, there may be various conflicts of interest, different expectations, diverse interpretations and lack of information, while the original conditions for adhering to the rules and norms, once adopted, may change in a turbulent market. Therefore, a corporate decision to comply or default once a firm has enrolled in a program is a complex matter. Firms have incentives to comply in the areas of climate change, although the climate issue in some of the cases dealt with is only one out of several parameters that form their reputation. In this context, Clapp and Thistlethwaite stress that measuring performance is a problem that remains to be solved, but it is an issue that must be built into the design of programs; indeed, all programs grapple with this issue. Those designing programs must anticipate what kind of problems will later arise, and, therefore, due diligence must be exerted. If programs have not established a clear and professional profile, it is really difficult to communicate to different publics when cases of non-compliance occur and to set out the measures that are appropriate to enhance compliance.

In forestry, examined by Bozzi, Cashore, Levin and McDermott, we find significant compliance among member firms of some programs; but here monitoring is also strong, firms are dedicated, and it is an area in which climate is of major importance. Some of the same features can be found in some programs in finance, but one should remember that the new programs analyzed are also specifically geared towards handling climate issues. In the broader government-initiated programs, compliance is less pronounced. In food, studied by Fuchs and Boll, and in cars, analyzed by Porter, affairs appear to be different because compliance among the limited number of member firms is still high, although the goals to be met in the programs are quite modest, climate being one issue among many. The formulation of broad-brush road maps, such as the Sustainable Mobility Project in the automobile case, can be a useful achievement of a private voluntary program, which may encourage firms to coordinate or otherwise alter their individual strategies without a compliance mechanism.

However, if programs selfishly care only about the performance of their own constituencies and isolate themselves from those that are not enrolled, they sidestep the free-rider problem. Indeed, non-compliance seems common among firms that have not joined the programs and statements are not followed up by practical action – as illustrated in forestry and finance, where there is a clear separation of firms in terms of commitment to programs.

The use of effective sanctioning mechanisms – strong and weak swords in the terminology of Hsueh and Prakash – shows the maturity of programs, as well as the determination to take care of this delicate issue. Not only are sanctioning mechanisms in place, but there is a spirit of preserving good faith and a desire to build upon those firms that have decided to do something about climate change already, although in their cases climate is only gradually developing as a core issue. Sanctions seem easier to implement in forestry than in food, but in both cases certification can be withdrawn; as well, the various public-driven programs have different degrees and practices of sanctioning non-compliance, while in the car industry's R&D collaboration and in the different regional programs in place here, there is a tradition of settling things in a diplomatic manner rather than expelling firms.

Multiple comparisons

This book proceeds from the view that there are many ways in which climate change problems can be mitigated and private voluntary programs can make a contribution to halt climate change in and across sectors. Some areas of business are cross-sectoral in nature and link up with other, more specific sectors. Energy producers, for instance, are serious emitters of greenhouse gases, and if problems are solved here, advances are likely to spill over into other areas. However, these other areas also have an impact on the level of energy production, and how energy is produced. Take, for instance, the car industry: both more efficient cars using traditional sources of energy and new hybrid or electric cars will have a tremendous impact on energy demand, and in these contexts private regulation is one of several tools that can be employed. It is also obvious that these other industries are not passively waiting for the energy producers to move first. They are themselves under tremendous pressure from competitors and customers to curtail emissions, bring new climate-friendly technologies to the market, and offer pioneer solutions.

Programs with a cross-sectoral orientation spring into existence, but they differ hugely in their organizational set-up. This book documents that a cross-sectoral field, such as the financial industries analyzed by Clapp and Thistlethwaite, has taken steps to mitigate climate change – steps that can have impacts on other sectors of business. In this case, key processes in the market stimulate the building of programs, but we also find some encouragement from governments and various private stakeholders. The climate issues are difficult to organize in relation to the financial industries, however, and many new initiatives have emerged, but they are characterized by an apparent lack of coordination of collective action.

By moving into programs that appeal to a multitude of sectors and large groups in business, these programs clearly can, in principle, be supported by multiple players, but multisectoral programs do not build on the same kind of basic understanding as that found in a typically cross-sectoral area, such as finance. These programs are less underpinned by underlying and existing organizational structures in business. Instead, as we have seen in the chapter by Hsueh and Prakash, government initiatives can be decisive in designing and running this kind of multisectoral program, and they compensate for the lack of large-scale business collective action. However, oversight is generally weak, and we should take a serious look at the engagement of public institutions, as their involvement is not always from a position of strength but rather from a place of weakness where possible interference with business authority is undesirable.

If we turn to specific sectors, we see that industries all have their own institutional properties, in which the various private voluntary programs are embedded, but certain key patterns are observable when they are compared (Begg, van der Woerd and Levy 2005). These patterns need continuous examination because conditions are changing in the turbulent field of climate policy and how it impacts on particular sectors. Indeed, such studies can serve as a further inspiration for analyzing industries not covered by this book.

Unlike the cross- and multisectoral programs, industry programs have the general advantage of working with specific issues of importance to comparatively smaller and more homogeneous groups of actors, and consequently a smaller part of the overall business community needs to be organized. These programs, however, build on a very uneven degree of support from existing organizations. Collective action theory has stressed the privileged role of small groups, with members sharing the same characteristics (Olson 1965), and we have already pointed to the role of this factor. From the standpoint of evolution theory, however, this is a case of "intraspecific competition", according to which the homogeneity of the actors constituting a group active in the same market is not always an advantage. On the contrary, strong rivalry is also likely to persist.[1] Industry programs, therefore, must further compete with unilateral initiatives of firms, especially in highly concentrated industries in which strong asymmetries prevail, and in which large firms have a potential to develop and communicate their own independent initiatives and position themselves as pioneers. Industry programs and single-firm programs, however, do not necessarily exclude each other. A thorough analysis is therefore needed to unravel which barriers and facilitators exist in specific industries.

In the industries reviewed here, one would expect that, being placed late in the production and value chain, the relatively proximity of firms to the end-consumers would benefit program building, because these countervailing groups would leverage firms to adopt stricter measures to halt climate change. This is only partly true, and this pattern applies mainly to forestry, which has made some early advances but also relies on public involvement, as the recent development demonstrates. In the food and car industries, this pressure is considerably smaller, and there is much ambivalence among both consumers and producers when it comes to abandoning certain lifestyles and to reforming production. The variation better lends itself to explanation if we consider that the environmental movement is a much stronger player than the consumer movement in relation to climate change politics, and that environmental concerns have more easily grown into climate change concerns.

There are many interesting features common to private voluntary programs, but such programs are also strikingly different – there is variation between cross- and multisectoral programs addressing a large part of the business community, and sectoral programs targeted at specific industries, and there are also different ways to arrange programs within these two subcategories. However, they all grapple with aligning private and public goals, they must all relate to other interested parties in their institutional environment, and they all have to solve problems at different stages of policy development and implementation. Taken together, the experiences from different private voluntary programs can be useful for scholars with a general interest in institutional innovation, its potentials and its pitfalls.

Ending and beginning

It is appropriate to end this book with an ambition to begin new research. Inspiration has been sought in different theories and the studies show that experiences with private voluntary programs are also relevant to different strands of research. It is not necessary to consider each of these in more detail than already discussed, but at a more general level we will briefly discuss some ramifications for future research into public and private regulation, into the study on politics and business, and into research on international affairs. Indeed, future studies on climate policy can benefit from all these research traditions.

Public regulation and private voluntary programs are not separate; public regulation includes creating a framework for private regulation and recognizing private initiatives, but there are also examples of a non-policy towards private efforts and a lack of coordination. This has been referred to as a "language problem" by key organizations in business (WBCSD 2009). Therefore, the many initiatives to foster regulation beyond the public sector in climate change politics cannot simply be conceived of as deregulation, as one might think from listening to the debate. This would require, e.g., a more directing role and more distinct principles.

On the private side, the public-good dimension is to varying degrees aligned with private goals, and it is a perennial challenge to bring these together, and the integration of a long-term perspective which is needed in climate policy is particularly demanding. From the perspective of business, policy is not simply delegated but often taken on by various private entities, and in such cases regulation is sometimes managed without dialog between public institutions and private actors. The centrality of public regulation varies, and a great amount of pragmatism prevails which leads to forms of cooperation and alliances with other available private parties that may contribute to consolidating an arrangement. In that sense various types of private arrangements can be conceived of as some of the "building blocks" in future climate policy (Falkner, Stephan and Vogler 2010).

The relation between politics and business is far from exclusively a matter of private inputs into traditional public policymaking, including governments and intergovernmental organizations. Key policymaking is also performed at the level of the business community, between business and other parties, and, of course, it also evolves in the terrain between business and public institutions. In these contexts it must be decided at which level business political action should be located: a mix of single firms, formal associations and alliances are activated, but at a time when climate policy is under development no particular and established model is in place. These different modes of action very much reflect the diversity of interests and strategies in business, and the many centrifugal tendencies lead to a multiplicity of programs, with public institutions often having little influence on correcting this diversity.

Finally, we need to draw some lessons for the study of international affairs. Many studies on climate policy focus attention on the role of states and intergovernmental organizations. More than anything else, perhaps, it is the stalemate and the slow progress at the level of governments, measured against the huge tasks facing us, that call for action beyond public authority. We have some exciting findings from the private sector that are definitely not fully developed but, nevertheless, give an indication of a potential possessed by private actors. Research into a number of intermediate institutions in and around the business community can give us a better guide to understanding initiatives that relate to but cannot be categorized as "state" or "market".

A successful grappling with these issues depends on keen advances in research and also on the real politics of climate change. Following up the attempts before and after the meager Copenhagen Accord reached at the COP 15 and the following COPs, we can envisage different scenarios. Will the lack of initiatives on the part of states and the non-fulfillment of international agreements water down the focus of business on climate change and inhibit the process of building private voluntary programs, or will this inaction on the part of public authority be taken as an opportunity by business to bring new programs to life and to improve those already in place? Or will renewed and concerted efforts of states and agreements at the intergovernmental level inspire business to follow up actions in areas where a general framework is created, and find novel solutions where these are still missing or vague, or will stronger public efforts instead be considered sufficient, leaving little space for business? These questions remain to be answered in the years ahead, but it is essential to bear in mind that private voluntary programs thus far have had many political, economic, technological and societal drivers that stimulate change, and, consequently, no single regulatory scenario is likely to prevail.

Note

1. This phenomenon has classically been observed in nature: "As species of the same genus have usually, though by no means invariably, some similarity in habits and constitution, and always in structure, the struggle will generally be more severe between species of the same genus, when they come into competition with each other, than between species of distinct genera" (Darwin 1859: 76).

REFERENCES

Begg, K., F. van der Woerd, and D. Levy, eds. 2005. *The Business of Climate Change: Corporate Responses to Kyoto.* Sheffield: Greenleaf Publishing.

Darwin, C. 1859. *The Origin of Species by Means of Natural Selection. Or The Preservation of Favoured Races in the Struggle for Life.* (First Edition). London: John Murray.

Delmas, M., and O. Young, eds. 2009. *Governance for the Environment: New Perspectives*. Cambridge, UK: Cambridge University Press.

Falkner, R., H. Stephan, and J. Vogler. 2010. "International Climate Policy after Copenhagen: Toward a 'Building Blocks' Approach." In *The Governance of Climate Change: Science, Politics and Ethics*, eds. D. Held, A. Fane-Hervey, and M. Theros, 202–222. Cambridge, UK: Polity Press.

Gunningham, N. 2007. "Reconfiguring Environmental Regulation: Next-Generation Policy Instruments." In *Industrial Innovation and Environmental*

Regulation: Developing Workable Solutions, eds. P. Saeed and B. Herbert-Copley, 200–232. Tokyo: United Nations University Press.

Hirschman, A. O. 1982. *Shifting Involvements: Private Interests and Public Action.* Princeton NJ: Princeton University Press.

IATA. 2009. *Airlines Present Climate Change Proposals to Heads of Governments.* Press Release No. 40, September 22, 2009. ⟨http://www.iata.org/pressroom/pr/2009-09-22-01.htm⟩; accessed September 26, 2009.

ICS. 2010. *Prevention of Air Pollution from Ships: Control of Greenhouse Gas Emissions from International Maritime Transport.* Submitted by the International Chamber of Shipping. January 10, 2010. ⟨http://www.marisec.org/news/Industry%20Submissions/ICS%20MEPC%20GHG%20Submission.pdf⟩; accessed January 26, 2010.

Meckling, J. 2011. "The Globalization of Carbon Trading: Transnational Business Coalitions in Climate Politics." *Global Environmental Politics* 11 (2): 26–50.

Olson, M. 1965. *The Logic of Collective Action: Public Goods and the Theory of Goods.* Cambridge MA: Harvard University Press.

Porter, T., and K. Ronit. 2010. "The Complexities of Global Business Authority and the Promise of Democracy." In *The Challenges of Global Business Authority: Democratic Renewal, Stalemate or Decay?* eds. T. Porter and K. Ronit, 279–309. Albany: State University of New York Press.

Scholte, J. A. 2002. "Civil Society and the Governance of Global Finance." In *Civil Society and Global Finance*, eds. J. A. Scholte with A. Schnabel, 11–32. London and New York: Routledge.

UNWTO and WTTC. 2009. *The Travel & Tourism Industry sets the Standard on Climate Change.* Joint Communiqué from UNWTO and WTTC, Madrid, Spain/Copenhagen, Denmark, London, UK, December 16, 2009; accessed January 26, 2010.

Vogel, D. 2008. "Private Global Business Regulation." *Annual Review of Political Science*, 11: 261–282.

WBCSD. 2009. *Governments Cannot Deliver Without Business, Says Stigson.* WBCSD News. ⟨http://www.wbcsd.org/Plugins/DocSearch/details.asp?DocTypeId=251&Object Id=MzY5NTc⟩; accessed December 30, 2009.

Index